Two week loan

Please return on or before the last
date stamped below.
Charges are made for late return.

IS 239/0799

INFORMATION SERVICES PO BOX 430, CARDIFF CF10 3XT

The Politics of Uncertainty:

Regulating Recombinant DNA Research in Britain

The Politics of Uncertainty:

Regulating Recombinant DNA Research in Britain

David Bennett, Peter Glasner and David Travis

Routledge & Kegan Paul

London, Boston and Henley

First published in 1986
by Routledge & Kegan Paul plc

14 Leicester Square, London WC2H 7PH, England

9 Park Street, Boston, Mass. 02108, USA and

Broadway House, Newtown Road,
Henley on Thames, Oxon RG9 1EN, England

Phototypeset in Linotron Times 10/11 pt
by Input Typesetting Ltd, London
and printed in Great Britain
by Billing & Sons Ltd, Worcester

Library of Congress Cataloging in Publication Data

Bennett, David J., 1937–

The politics of uncertainty.
Bibliography: p.
Includes index.
1. Recombinant DNA—Research—Social aspects—Great
Britain.   2. Recombinant DNA—Research—Law and legis-
lation—Great Britain.   I. Glasner, Peter E.   II. Travis,
G. David L. III. Title.
QH442.B46   1986   363.1'9   85–18373

British Library CIP data also available

ISBN 0–7102–0503–1

Contents

Foreword

The Williams Report, published in August 1976, included a proposal that 'those who plan to work in the field of genetic manipulation [be required] to submit their experimental protocols to a central advisory group'. The British government accepted the proposal with rare alacrity, and it became the responsibility of the Secretary of State for Education and Science, Mrs Shirley Williams, to decide upon the membership of the Genetic Manipulation Advisory Group (GMAG).

Mrs Williams made it clear, by prompt action in September and October, that she was alive to the urgency of an acceptable procedure, but then had the audacity to compose the Group of such diverse elements that it hardly seemed possible that it could succeed. The members included representatives of science, the universities, industry, trade unions and the general public; there were experts in genetic engineering, the handling of dangerous pathogens and radiation hazards, also a medical practitioner, a psychiatrist, a science journalist and a philosopher and ethicist. Since the Group numbered sixteen, most members served in more than one capacity. The background was discouraging: scientists generally were hostile, trade unionists deeply suspicious, industry almost paralytic with caution, whilst the media strove in vain to incite the hysteria then prevalent in the USA.

But GMAG worked, calmly and constructively, throughout a period when public reassurance was essential. There were perilous moments of polarisation, of the crystallisation of stereotyped opinions. Perhaps the most vital quality of membership was a sense of humour, but there was a healthy readiness to accept each other as people and not as stock representatives of big science, management or employees. The technicalities may have remained beyond comprehension but the elements of health and safety were made clear, and each member found an important personal role.

GMAG worked its way out of a job, out of the need for continued public reassurance. Quasi-official bodies are now unpopular with government but I believe, as the first chairman of GMAG, that it is in the national (and international) interest that this experimental group should not be forgotten. The world presents all too many problems on which the general public needs and deserves informed reassurance. At least one such national group appears to be essential; liaison with other national groups no less important. If the example of GMAG proves fruitful, it will be one more significant way in which genetic engineering gives hope to a desperately anxious world. I hope this book, prepared with such care, points compellingly in that direction.

Sir Gordon Wolstenholme, OBE, LLD, FRCP
Chairman of the *Genetic Manipulation Advisory Group, 1976–8*

Acknowledgments

We would like to thank all those concerned with genetic manipulation and with GMAG who so generously gave of their time and of their experiences to this project. We would particularly like to thank Sydney Brenner, Donna Haber, Noel Parry, Jerry Ravetz, Charles Weiner, Bob Williamson and Ed Yoxen for their specific contributions, and John Subak-Sharpe for his detailed comments on an earlier draft. We would also like to thank the Polytechnic of North London for housing the project and funding the research upon which this book is based.

Glossary of acronyms

ABPI	Association of the British Pharmaceutical Industry
ABRC	Advisory Board for the Research Councils
ACGM	Advisory Committee on Genetic Manipulation
ARC	Agricultural Research Council
ASTMS	Association of Scientific, Technical and Managerial Staffs
AUT	Association of University Teachers
BSSRS	British Society for Social Responsibility in Science
CACS	Central Advice and Control Service
CAG	Central Advisory Group
CAS	Central Advisory Service
CBI	Confederation of British Industry
CERB	(USA) Cambridge Experimentation Review Board
COGENE	Committee on Genetic Experimentation of the International Council of Scientific Unions
CVCP	Committee of Vice-Chancellors and Principals
DES	Department of Education and Science
DHEW	(USA) Department of Health, Education and Welfare
DHSS	Department of Health and Social Security
DNA	Deoxyribonucleic acid
DPAG	Dangerous Pathogens Advisory Group
EEC	European Economic Community
EMBO	European Molecular Biology Organisation
GMAG	Genetic Manipulation Advisory Group

Glossary of acronyms

GMP	Good Microbiological Practice
GMUG	Genetic Manipulation Users Group
HASAWA	Health and Safety at Work Act
HSC	Health and Safety Commission
HSE	Health and Safety Executive
IPCS	Institute of Professional Civil Servants
MAFF	Ministry of Agriculture, Fisheries and Food
MRC	Medical Research Council
NALGO	National and Local Government Officers' Association
NAS	(USA) National Academy of Sciences
NIH	(USA) National Institutes of Health
QUANGO	Quasi-Autonomous Non-Governmental Organisation
RAC	(USA) Recombinant DNA Advisory Committee
TUC	Trades Union Congress

1
Introduction

Genetic manipulation, the joining of hereditary material from disparate sources and its reinsertion into an organism, has recently completed its first decade. During this time, it has developed from the initial discoveries of the technical procedures which made it possible to its widespread use in many areas of biological research, and to the early stages of its application in a wide range of biological, medical and agricultural technologies. Its development has been attended by controversy within science and in society at large.

Almost from the outset scientists recognised the techniques as having far-reaching scientific and technological benefits, but also the potential for creating hazards both to humans and to the environment. This was especially the case since the bacterium most often used in experiments (*Escherichia coli*) is a normal inhabitant of the human intestine. Once attention had been drawn to the potential for hazard associated with recombinant DNA experiments, a call for a voluntary moratorium on existing research was made, in 1974, by a group of leading American scientists. By publicising their concerns, a debate was initiated about the regulation of genetic manipulation experiments which quickly embraced the wider social and ethical issues involved, and spread beyond the borders of science into the public and political arenas.

In Britain, the response to this call for a temporary halt to experiments was the establishment of a Working Party under Lord

Ashby, to report on the potential risks and benefits of genetic manipulation. It recommended that research should be allowed to continue because of the likely benefits, but should be subject to rigorous safeguards provided by a voluntary code of practice. A further Working Party was set up under the chairmanship of Sir Robert Williams in 1975 to develop these recommendations. This led a year or so later to the establishment of the Genetic Manipulation Advisory Group (GMAG) with the task of vetting proposals for experiments using the new techniques. GMAG was an interesting innovation in science policy because it contained representatives of the public interest, and of the interests of employers and employees, as well as scientific and medical experts. Meanwhile, in the United States, the National Institutes of Health, which is responsible for most of the funding for academic research in the biomedical field, laid down detailed guidelines which scientists were expected to follow, although these did not apply to industrial laboratories. The interpretation and regulation of the guidelines were left at local level, thereby paving the way for stormy confrontations between various interest groups including concerned scientists, local politicians, and the general public. The events in the United States are now quite well documented,[1] but this is not so for the development and operation of genetic manipulation regulations in Great Britain.[2] The intention of this book is to provide a sociologically informed account of events in the United Kingdom since the early 1970s.

   The book is based upon an extensive research project begun in 1979 (Bennett, Glasner and Travis, 1984), which included many interviews with scientists, politicians and others associated with regulating recombinant DNA research in Britain, and a detailed analysis of a wide range of public and private documents, archival and secondary material. The main official documents used were the Reports of the Genetic Manipulation Advisory Group, which include lists of members, sub-committees, participating laboratories, discussions of the nature and level of controls, numbers of proposed experiments, and reports on the operation of the Group over the period covered. In addition, there were also House of Commons Select Committee and Working Party Reports, GMAG Notes to scientists, Health and Safety Executive documents covering laboratory procedures and various documents prepared by the EEC, European Molecular Biology Organisation, and the United States Government and its Offices. Most of the detailed information regarding the establishment and operation of GMAG came from semi-structured interviews with GMAG members, including chairmen, civil servants, scientific and medical experts, trade unionists, employers, politicians, lay representatives of the public interest, and members of the Health and Safety Executive.[3] The authors were given access to the collection of private papers of Dr Sydney Brenner, a leading scientist and participant in British and American developments, and Director of the MRC Laboratory of Molecular Biology, Cambridge. Other valuable material was provided by various respondents and the Recombinant DNA Oral History Collection at the Massachusetts Institute of Technology. A useful body of reliable secondary material is to be found in *Nature*, *Science* and *The New Scientist*, which primarily document events as they occurred, and journals such as *Minerva*, *Scientific American* and *The Bulletin of the Atomic Scientists*, which contain accounts of scientific and public issues of a more reflective nature. There is also a large number of news items, feature articles and editorials in the serious popular press, as well as many books and articles generated by commentators on the debate from a variety of backgrounds.

The research, therefore, provides the basis for the first comprehensive account and analysis of events in Britain. In doing so, it utilises the general perspective of what has come to be called the post-Kuhnian sociology of science which is discussed briefly below. It also provides a useful basis for comparative analysis, not only of how scientists handle new technology, but also of how science policy is developed and applied, and can therefore contribute to filling the gap between the social studies of science and science policy studies (see, for example, Spiegel-Rösing, 1977:29).

The relationship between scientists and technologists to political power has inevitably altered in modern times (Lakoff, 1977:356). Leading and prominent scientists have now acquired unprecedented access to the decision making fora of contemporary society. But, because of the growth of 'big science', with its heavy dependence on public subvention, government can now influence and direct technological advance in ways previously not thought feasible. In addition, many scientists now feel a degree of social responsibility for the use to which their work is put, and a duty to participate in the political process which determines this.

One of the techniques developed by science policy analysts to aid governmental decision making in the area of technological development is risk assessment research. This attempts to investigate a range of objective outcomes to a given technological advance, and to provide for a rational choice in the light of given goals and resources (Wynne, 1975). In effect, it aims to restore neutrality, and assumes that a scientific consensus can be objectively created and that scientific knowledge can legitimise decisions arrived at concerning the goals of research. It often suffers (Johnston, 1980) from claiming certainty of knowledge, and monopoly over expertise, while occasionally seeing its role as educating society into the 'correct' policy option (Wynne, 1980). Technology or risk assessment research also assumes a consensus of views within society at large about the goals and aspirations of scientific advance. Hence it functions as something of a 'rhetorical smokescreen' (Wynne, 1975:112) veiling both the differences within science and the complexities of the wider political process. One aim of this analysis of the controversy surrounding the use of genetic manipulation techniques in Britain is to clarify some of these issues. Many studies of science policy have, until recently, treated the cognitive aspects of science as a 'black box' (Whitley, 1972), thereby only providing a partial view of these important developments.

Alternative approaches come from recent advances in the sociology of science (Hesse, 1980), which start from the premise that the approaches to the history of science previously labelled 'internal' or 'rational' and 'external' or 'social' are really complementary (Johnston, 1976). They go on to assume a certain degree of relativism concerning the status of the scientist's knowledge about the world, based in part on the contributions of Thomas Kuhn (1970).[4] And, perhaps more importantly, they tend to look to the controversial rather than the normal processes of science so as to see more clearly how the structure of science operates. This approach contrasts with earlier, functionalist models of the social system of science (Merton, 1942; Storer, 1966) as Mulkay (1980:11) notes:

> In contrast, researchers in Europe emphasized what they saw to be important and analytically provocative differences of perspective. Mertonian analysis was seen as involving a definite separation of cognitive and social processes in science, in accordance with the traditional assumptions of the sociology of knowledge. . . . Kuhn's analysis seemed to raise the possibility that the cognitive, social and moral aspects of science were linked together in a much more complex way than had previously been assumed; indeed, that scientists' technical conclusions were established partly by means of social commitments and that sociologists need no longer presume that the cultural products of science had been validated by the application of universal criteria which sociologists themselves had to accept as analytically unproblematic.

The range of empirical work covered by these new approaches to the sociology of science is by now sufficient to establish a strong British and European tradition, although not necessarily a unified one.[5] The approaches are echoed in this book, particularly in the emphasis on a relatively microsociological style of analysis. Scientific arguments concerning the possible risks of genetic manipulation are discussed, but only where relevant as part of policy decisions. Few, if any, of the issues in the controversy have been purely scientific and, after the initial explosion of concern, there has been an increasingly strong scientific consensus that there are no specific risks in genetic manipulation techniques as such. As Yoxen (1979a:227) has argued, the public unanimity has effectively blanketed any latent controversy, and closed off points of access. By focusing both on the cognitive aspects of a controversy as well as the broader, social influences, within the framework of contemporary events, these approaches, like this book, provide greater insight into the processes of scientific development.

The first part of the book provides the reader with a fairly detailed analysis of the development and operation of genetic manipulation regulations in Britain. It begins with a brief recapitulation of events in the United States of America and recounts the developments in Britain, including the Working Parties chaired by Ashby and Williams. It analyses the reactions of the scientists to the proposals for guidelines on research, and the effect of wider social issues on the decisions made. This is followed by a close look at the early days of the Genetic Manipulation Advisory Group, the roles of the chairmen, secretariat and other groups, the establishment of a routine for dealing with applications, and the overall balance which developed between different, and occasionally conflicting interests. It continues with an analysis of the initial basis upon which proposed experiments were categorised, and the need for changes to keep in line with advances in scientific knowledge and industrial application. The routinisation of the processing of applications from laboratories is discussed, concluding with the change in the status and composition of GMAG in 1984 from an independent quango to becoming part of the Health and Safety Executive.

The book continues with a discussion of the crucial issues raised during the life of the Genetic Manipulation Advisory Group, the scientific basis for assessing potential hazards. This is analysed in detail to show how the phylogenetic basis of the original proposals was replaced by a probabilistically based scheme, which effectively reduced the number of experiments which fell into higher containment categories. The penultimate chapter outlines the background to the moratorium, both in Britain and America, where ideological, social and economic factors all contributed to making genetic manipulation a focus of concern. It goes on to discuss the move from uncertainty about hazards to calculations about risks, and further to the routinisation of the risk assessment process. In particular, this chapter highlights the problems concerning expert knowledge faced by the lay representatives of the public interest, and their relationships with the members of the other constituencies represented on GMAG. In conclusion, there is an evaluation of the success of this British approach to regulating genetic manipulation experiments in the light of the translation of GMAG from a quango to a committee of the Health and Safety Executive.

The overall balance of the book is towards a discussion and analysis of the events associated with the establishment, operation, and eventual demise of the Genetic Manipulation Advisory Group. It is an attempt to provide an informed, accurate, though necessarily partial, record of what must be considered a unique experiment in the process of regulating scientific research.

2
The development
of a British policy

The development of a British policy on the regulation of experiments with recombinant DNA was closely linked with the events which took place in the United States in 1973 and 1974. This chapter begins with a brief review of these events, before moving, in more detail, to the particular sequence of events leading to the establishment of guidelines for research in Britain. As noted in the Introduction, much of the American material has been collected and analysed in some detail, particularly in Krimsky's (1982) lengthy social history of the controversy, while Watson and Tooze (1981) have reproduced many of the key documents. The first part of the chapter does not, therefore, set out to be an exhaustive account. It is intended to help the reader put events which took place in Britain into context.

The American experience

The promise of the genetic manipulation techniques and the question of possible risks were first raised for general discussion in June 1973 at the Gordon Conference on Nucleic Acids in New Hampton, New Hampshire. One of the sessions towards the end of the conference was devoted to the new genetic manipulation techniques. After a fairly short discussion the participants, some of whom were the pioneers in the research, voted that a letter should be sent to the President of the USA National Academy of Sciences expressing their concerns. These were mainly about the

7

possibility of creating new hybrid viruses with unpredictable biological properties, particularly oncogenic (cancer-inducing) viruses. The letter, which was published in *Science*, a journal with a very wide circulation, and in the *Proceedings of the National Academy of Sciences*, in July 1973 over the signatures of the co-chairmen of the conference session, suggested that the Academy should 'consider this problem and recommend specific actions or guidelines' (Singer and Söll, 1973). The Academy responded by appointing a study panel of scientists who were actively engaged in recombinant DNA research and chaired by Dr Paul Berg, Professor of Biochemistry at the Stanford University Medical Center, to examine the question.

After nearly a year of discussion, the panel issued its report and a second letter summarising its findings was simultaneously published in *Science*, *Nature* and the *Proceedings of the National Academy of Sciences* in July 1974. This letter, which came to be known as the 'Berg letter', called for a 'moratorium', or defer-ment, of those experiments which involved the joining of DNA segments from oncogenic or other animal viruses, or the insertion of certain types of genes into bacterial plasmids (small circular DNA molecules capable of replicating themselves many-fold within the bacterium):

> First, and most important, that until the potential hazards of such recombinant DNA molecules have been better evaluated or until adequate methods are developed for preventing their spread, scientists throughout the world join with members of this committee in deferring . . . . [certain] experiments. (Berg *et al.*, 1974)

It was also suggested that so-called 'shotgun' experiments linking animal DNA to bacterial plasmids in a random manner should be 'carefully weighed'. The letter called for the National Institutes of Health (NIH)[1] to establish an Advisory Committee for 'devising guidelines to be followed by investigators working with potentially hazardous recombinant molecules', and for an international conference of experts to discuss the issues.

This letter, together with that from the Gordon Conference participants, and their wide publication, initiated a series of events which converted the early concerns into a major social and scientific issue. The 'Berg letter' established certain important precedents which were to become central to succeeding events. Thus an international meeting was recommended since the hazards, should they eventuate, would not be constrained by national boundaries. The letter also called for the establishment of government regulatory bodies so that the interests of the general public could be represented, and proposed that the discussions be open and well publicised. The signatories of the 'Berg letter' have since indicated that their motives included, in varying part, genuine concern, the wish to demonstrate the social responsibility of biological scientists by comparison with the actions of the early atomic physicists, and the desire to stake a prior claim in the field (Watson, 1977).

As a result of these letters the NIH established a Recombinant DNA Advisory Committee in October 1974, and the National Academy of Sciences sponsored an international conference at Asilomar, California in February 1975. A great deal of publicity was generated by these events which now moved the debate into the public arena. It is important to recall that the issue arose at a time when others which were to exacerbate it were in the forefront of public attention in the United States. The Watergate episode had suspended trust in government institutions. The radical university movement had lost its focus with the end of the Vietnam war. The honeymoon between pure science and government had ended. Resources had become limiting, and were generally perceived to be so, even in the United States. Mistrust of scientists had grown, many seeing them as self-interested at the expense of the public and supported by the NIH, itself composed of a peer group with similar motivation.[2] Of particular relevance to genetic manipulation, general awareness of the harmful effects of many technologies, especially those having hazards for health or leading to environmental pollution, had become an important political issue. Thus the climate was ripe for the conversion of the concern about the conjectured risks of genetic manipulation into a major controversy when they became publicised. The controversy was fuelled by the pronouncements of a few prominent or vocal scientists who voiced strong opinions about the possible apocalyptic consequences of genetic manipulation, and sensational capital was made by sections of the popular media out of the potentials in the debate for science fiction-like accounts and predictions of horrific dimensions.[3] However, the majority of the concern was serious and strongly felt.

The concern of sceptics and opponents of genetic manipulation was sharply articulated, for example, by Robert Sinsheimer, chairman of the Division of Biology at the California Institute of Technology. He asked, 'Do we want to assume the basic responsibility for life on this planet? To develop new living forms for our own purposes? Shall we take into our own hands our own future evolution?' (Sinsheimer, 1977:138). Since the questions included

9

such concepts as responsibility, purpose, and control of the future genetic composition of human beings and other organisms, they clearly involved considerations well beyond science. Sinsheimer entered yet another controversial area by wondering whether 'there are certain matters best left unknown, at least for a time' (Sinsheimer, 1977:139), thus questioning the fundamental premise that the growth of knowledge is the engine of other values, and is not driven by them. Chargaff, a pioneer in the deciphering of the genetic code, epitomised this general viewpoint by maintaining, 'My generation, or perhaps the one preceding mine, has been the first to engage, under the leadership of the exact sciences, in a destructive colonial warfare against nature. The future will curse us for it' (Chargaff, 1977:139). Sinsheimer and Chargaff, together with certain philosophers, historians and sociologists of science, were clearly suggesting that the possible consequences of knowledge must be consciously included in decisions about the directions of the search for knowledge itself.

However, the National Academy of Sciences (NAS) committee, which organised the Asilomar conference at the behest of scientists engaged in genetic manipulation research, tended to ignore these wider social issues, preferring to define the issues in terms of producing a set of technical safety procedures to contain any potentially hazardous organisms or material. This was a stage in the evolution of what Wright (1978) has called the 'technological paradigm'. In general, diffuse social, ethical and political aspects of the debate were progressively eased out and the situation came to be defined in the more tractable terms of the technicalities of containment. Thus the NAS recommended that the Director of the National Institutes of Health establish an advisory committee to evaluate potential hazards of the research and devise guidelines to be followed by research workers. The NIH Recombinant DNA Molecule Program Advisory Committee (which later became the NIH Recombinant DNA Advisory Committee) was charged with the remit of advising the Secretary of the Department of Health, Education and Welfare, the Assistant Secretary of Health, and the Director of NIH:

> concerning a program (a) for the evaluation of potential biological and ecological hazards of various types, (b) for developing procedures which will minimize the spread of such molecules within human and other populations, and (c) for devising guidelines to be followed by investigators working with potentially hazardous recombinants. (NIH, 1977)

The first meeting of the committee was held on 28 February and

1 March 1975 in San Francisco, immediately after the Asilomar conference, and included twelve members, all scientists, with less than half of them actively employing the recombinant DNA techniques in their research, and the others more or less familiar with the subject (Grimwade, 1977). As the work proceeded more members were added, some representing important but underrepresented scientific areas such as the plant sciences, some to represent pressure groups such as Science for People, and some to represent the areas of ethics, law and government. Unfortunately, the area that was the main concern of the committee, that is to say the epidemiology of pathogenic organisms, such as might be produced by genetic manipulation experimentation, was underrepresented, especially after the early resignation of Falkow of the University of Washington.[4] Knowledge about the epidemiology of infectious diseases, which would have supplied information about the possibility of a host organism into which foreign DNA had been introduced infecting humans and spreading, in fact largely existed and could have been provided by epidemiologists, medical microbiologists and immunologists. The deficiency, which also characterised the composition of the Gordon Conference and signatories of the 'Berg letter', would later come to be seen as having played a significant part in allowing the early scientific estimates of the potential risks likely to be involved in the research to be overemphasised.

Meanwhile, the international conference had taken place at the Asilomar Conference Center in California on 24–27 February 1975, and published its 'Summary Statement' in *Science*, *Nature* and the *Proceedings of the National Academy of Sciences* (Berg *et al.*, 1975). The meeting was sponsored by the National Academy of Sciences and supported by the National Institutes of Health and the National Science Foundation. The Organising Committee consisted of five eminent scientists in the field, the chairman Paul Berg, David Baltimore, Sydney Brenner, Richard O. Roblin and Maxine Singer. These scientists chose the other participants on the basis of being 'presently engaged in or contemplating researches with recombinant DNA molecules' or 'experts who could provide special information and insight to the question of assessing and dealing with the potential hazards of such work'.[5] Some 150 people attended, including fifty-two scientists from outside the USA, sixteen press representatives, and four lawyers. The meeting was organised 'to review scientific progress in research on recombinant DNA molecules and to discuss appropriate ways to deal with the potential hazards of this work'. Of particular concern was whether the moratorium on certain areas of work called for in the 'Berg letter' should end and, if so, how research could proceed with minimal risks to research workers, the public, and to animals and plants in the environment.

In the 'Summary Statement' the principles taken as fundamental were:

(1) that containment be made an essential consideration in experimental design and,
(2) that the effectiveness of the containment should match, as closely as possible, the estimated risk. (Berg *et al.*, 1975)

It was accepted that the estimation of risks would be difficult given the paucity of information available at that time, and that it would have to be 'intuitive'[6] at first. But it was hoped that as a body of knowledge was built up the matching of potential hazards with the appropriate level of containment would improve and that it could be applied consistently. Four types of containment were proposed corresponding to four levels of possible risk.

(1) Minimal risk – for experiments in which the hazards could be accurately assessed and would be expected to be minimal. Containment would consist of such measures as no eating, drinking or smoking in the laboratory, wearing laboratory coats, the use of cotton-plugged or mechanical pipettes and prompt disinfection of contaminated materials. (Such precautions – 'good microbiological practice' – were already standard in, for example, laboratories in hospitals.)
(2) Low risk – where it was expected that novel organisms could be generated but that the recombinant DNA could not alter the ecological behaviour of the host, increase its pathogenicity, or prevent effective treatment of resulting infections. Mouth pipetting would be prohibited, biological safety cabinets would be used for procedures producing aerosols and access limited to laboratory personnel. As safer biological vectors (means of inserting the new, foreign DNA into the host) and host organisms became available they should be used. (This latter type of precaution became known as biological containment in distinction to the other physical containment procedures.)
(3) Moderate risk – for experiments where there would be the probability of generating an organism with significant potential for pathogenicity or ecological disruption. In addition to the previous measures for containment, hermetically sealed safety cabinets would be used for all operations, laboratories would be maintained at negative air pressure

relative to other rooms and corridors, and vacuum lines used in experiments would be protected by filters. Experiments would only be done with vectors and hosts that had an appreciably impaired capacity to survive or multiply outside the laboratory.

(4) High risk – for experiments in which the potential for ecological disruption or pathogenicity could be severe to laboratory workers or the public. This type of facility, similar to those designed to contain highly infectious organisms, would be isolated by air locks, have negative air pressure and require clothing changes and showers for personnel on leaving. All exhaust air and wastes would be treated to inactivate or remove any biological contaminants. Additionally, only vectors and hosts whose growth could be confined to the laboratory could be used. (Such *physical* containment procedures are normally used in, for example, research on highly pathogenic microorganisms and in microbiological warfare establishments.)

It was proposed that certain types of experiments be deferred since they presented such possibly serious dangers that they should not be carried out with available host vector systems and containment facilities. These included the construction of drug resistant,[7] pathogenic or toxin-producing organisms, the introduction into bacterial cells of all, or part of, the DNA from viruses known to cause cancer in animals or humans, and, finally, large-scale experiments using more than ten litres of culture.[8]

It was anticipated that research in the area would develop very quickly, and that it would be applied to very many biological problems, though the particular areas could not be foreseen. Therefore, it was held to be essential that there should be a programme of continual reassessment, probably by means of a series of annual international workshops. High priority should also be given to research to improve and evaluate the containment effectiveness of new and existing host-vector systems.

The 'Summary Statement' concluded by stressing the scarcity of information then available about the likely risks, and the necessity to mount experimental programmes to assess them, particularly before large-scale applications of the technology were attempted.[9]

It is important to note that even at this stage the strong links which existed between research workers in the field on both sides of the Atlantic had led to continuous interchange of information, not only about scientific developments but also concerning their apprehensions about the possible risks involved and the likely political implications. Dr Sydney Brenner of the MRC Laboratory of Molecular Biology, Cambridge (UK) was one of the five organisers of the Asilomar conference, and later maintained that the UK Ashby Working Party report which had been recently published, and which is further discussed in chapter 3, had an important influence on the Asilomar deliberations.

At the first meeting of the NIH Recombinant DNA Advisory Committee, held immediately after the Asilomar conference, it was recommended that NIH use the Asilomar guidelines for research until the Committee had an opportunity to elaborate more specific ones. Due to the considerable political pressures then bearing on it and the consequent sense of urgency, the Committee did not address itself to its primary assignments – 'the evaluation of potential biological and ecological hazards of DNA recombinants', but tacitly assumed that valid reasons existed for 'developing procedures which will minimize the spread of such molecules within human and other populations, and for devising guidelines' (NIH, 1977).

A subcommittee consisting of David Hogness as chairman, Ernest Chu, Donald Helinski and Waclaw Szybalski drew up the first draft of the guidelines on the basis of the original Asilomar statement. This draft was seen as a compromise between the divergent beliefs of the subcommittee members on the relative importance of the biological and political dangers, and what the response to them should be. It emerged as a cautious document, believed by the subcommittee to be necessary in order to allay apprehensions and to deflect any moves by the American government to impose regulations or legislation. The NIH Committee, taking the draft prepared by the Hogness subcommittee, developed a set of guidelines at its third meeting at Woods Hole, Massachusetts in July 1975. The 'Woods Hole draft', which tended to downgrade the previous conservative features, was circulated, but only to be greeted by critics some of whom felt that the revised guidelines were too lax, and others that they were too restrictive. A new subcommittee, chaired by Dr Elizabeth Kutter, was appointed to devise revisions of the guidelines. For a fourth meeting of the NIH Committee held at La Jolla, California on 4–5 December 1975 a 'Variorum Edition' (after Furness's Variorum Edition of Shakespeare) had been prepared in which the Hogness, Woods Hole and Kutter versions of the guidelines were compared in detail. The committee voted item by item for its preference between the three versions and, in many cases, added new material. The result was the 'NIH Proposed Guidelines for

Research Involving Recombinant DNA Molecules' which was referred to the NIH Director, Dr Donald Fredrickson, for final decision. He convened a public meeting on 9–10 February 1976 to allow further public scrutiny and review. Following these deliberations, the first official version of the guidelines was published on 7 July 1976 (Federal Register, 1976). They emerged as a conservative set of regulations designed to allay general apprehensions.

To a considerable extent they were successful in that aim, but, in the view of some scientists, at the expense of serious deleterious consequences for research. These included substantial delays in research which required intermediate categories of containment, and virtual prohibition of research requiring the higher containment categories, until laboratories could be built and strains developed. Moreover, a great deal of time and effort was required of both scientists making applications and committee members in assessing them, whilst research funds which could have been devoted to other uses, as it transpired, were used in designing and building high containment laboratories. Such views depend importantly on the prior identification of the issues surrounding recombinant DNA techniques as 'technical' safety issues. The more politically sensitive scientists recognised that there were political and social realities constraining what could be done, as well as their more familiar biological reality, and, of course, such views were also part of attempts to change the political reality of the situation.

As the experimental hazard potential came to be seen by scientists as progressively more unlikely, and the scientific basis of the guidelines came to be widely seen as weak, they tended to fall into disrespect. In fact, the Recombinant DNA Advisory Committee started revision of the guidelines soon after their initial publication, producing a new and much more specific set of proposals in September 1977 (Federal Register, 1977).

The American guidelines for genetic manipulation research have been 'voluntary' in that there is no legal requirement (at a Federal level)[10] to follow them. A number of proposals for Federal legislation were made, but these were successfully defused by scientists' lobbying efforts. Observance of the guidelines was, however, a condition of the receipt of government research grants, and this seems to have been a powerful constraint – certainly there are very few recorded cases of infraction of the guidelines. Industrial researchers were expected to observe the NIH guidelines, but there was no mechanism of ensuring compliance.

Most countries which support genetic manipulation research

have adopted some form of guidelines more or less closely related to those of the NIH. The only major alternative to this system has been that of the British Genetic Manipulation Advisory Group.

The British response

In outline, the first visible response was the establishment of a Working Party under Lord Ashby to report on the potential benefits and hazards of the new techniques in 1974 (Ashby, 1975:2). The Working Party's conclusion was that because of the likely benefits, experimentation should be allowed to go ahead, subject to rigorous safeguards (Ashby, 1975:15). This would entail a voluntary system of control and the provision of a flexible code of practice to guide researchers (Ashby, 1975:13). In August 1975, the Secretary of State for Education and Science set up a further Working Party under the chairmanship of Sir Robert Williams to give substance to these recommendations. The deliberations of the Williams Working Party led to the establishment of the Genetic Manipulation Advisory Group (GMAG) in the autumn of 1976.

Looking at the general implications of British (and American) policy formation, several commentators (Wright, 1978; Yoxen, 1979a; Krimsky, 1982) have argued that the process was one whereby the wider social and ethical issues which began to emerge early in the debate ('What are the social implications of the new technology?'; 'Should we allow the research to go ahead?') were successively ruled out of court in favour of the more narrow and technical issues of risk assessment ('What precautions are necessary to reduce the risks to acceptable proportions?'). The Ashby Working Party was asked to 'make an assessment of the potential benefits and potential hazards' (Ashby, 1975:iv), and pointed out early in its report that it was not 'set up to make ethical judgements about the use of the techniques' (Ashby, 1975:3). This can be read as part of the continuing process of establishing the ground rules – that the regulation of recombinant DNA research was basically a technical safety matter. As noted earlier, Wright (1978:1402) has called this the establishment of the 'technological paradigm'. The evolution of this particular variant of the 'technical fix' will be discussed in the following pages.

On the same day in July 1974 that the 'Berg letter' was published, Sir John Gray, Secretary to the Medical Research Council (MRC), circulated relevant laboratories asking, in effect, that its provisions should be observed. Also on the same day the Advisory Board for the Research Councils (ABRC) met to establish the Ashby Working Party.

Drafts of the 'Berg letter' had been circulating in Europe for some time and an early copy brought back from America had been sent to the MRC on 18 June by Professor John Subak-Sharpe, Director of the MRC Virology Unit at the University of Glasgow.[11] Subak-Sharpe was anxious that the MRC should be aware of the impending call for the moratorium, and should consider its response, especially as the new techniques would be of great economic importance, in addition to their obvious scientific value. He felt that these attractions might lead those with limited relevant scientific experience of handling potentially dangerous organisms – and hence little appreciation of the possible hazards and required precautions – to enter the field.

Comparatively little is known about the discussions within the MRC (or other research councils) and the senior scientific establishment concerning the appropriate form of response to the 'Berg letter'. It does appear, however, that at the higher levels the MRC was relatively more convinced of the necessity for a temporary moratorium than were similar levels within the Agricultural Research Council (ARC).[12] The ARC seems to have doubted the scientific grounds for the claimed possible risks. As it happened, the ARC was not sponsoring any research which involved the use of genetic manipulation techniques, nor was there any likelihood that it would do so in the immediate future. On the other hand, MRC-supported researchers had a much closer interest in the new techniques, and perhaps for that reason there was a general tendency for the ARC to defer to the MRC view.

The MRC was responsible for research units which were already beginning to use recombinant DNA techniques, though these were not the categories banned or questioned in the 'Berg letter'. Nevertheless, research programmes funded by the MRC would soon reach the stage where they could use to scientific advantage the questioned though less potentially hazardous of the new techniques. There was, therefore, a need to set in motion discussions that would lead to some kind of mechanism for deciding which categories of experiment could be performed. One early MRC plan involved the establishment of an internal organisation. In an interview with Professor Charles Weiner of the MIT Oral History Program on Recombinant DNA, Brenner stated:

> At [the] Cold Spring Harbour [conference] many people heard about recombinant DNA and came back and said that we must get involved in this work, but it's dangerous, potentially

dangerous . . . there were some discussions that in fact
maybe the best thing would be for the MRC to set up some
organisation [to let the research go ahead] . . . We already
had on paper a number of people whom one would have got
together. As it happened then, it was decided that this was
beyond the MRC. (Brenner, 1975:37)

The Advisory Board for the Research Councils which established
the Ashby Working Party is, like the MRC, effectively under the
control of the Department of Education and Science (DES). The
recombinant DNA issue involved matters relating to public health
and the Department of Health and Social Security (DHSS) would
have been consulted before the Ashby Working Party was set up,
and must have given at least tacit approval to the scheme.

The decision to reject the idea of an internal MRC committee
in favour of the ABRC/Ashby approach was partly influenced by
the recognition that if separate actions were taken by the research
councils, the most restrictive line would almost inevitably become
the accepted standard. In any case, different approaches would
suggest differences of opinion, and this could well undermine
public confidence. It was stressed in several interviews that the
need to provide visible assurance that the issues had been given
full consideration was of major importance; and it was felt that
the relative independence of the ABRC/Ashby approach would
command greater legitimacy. But, whatever the strength of the
feeling that there was a need to reassure the public, this did not
run as far as the establishment of an independent body to consider
the wider issues surrounding genetic manipulation. In this regard
Wright (1978:1400) has emphasised that decision making about
recombinant DNA took place within institutions concerned with
the promotion of biomedical research, and, in a similar vein,
Yoxen (1979a:228) has observed that when the Ashby Working
Party was established it was with a 'narrow spectrum of scientific
expertise', although from a range of biological disciplines.

The particular context in which the processes of policy formation took place is, of course, an important aspect of the analysis of
the outcome. Although the MRC did become the leading research
council in the development of policy, it could not act in isolation.
Also, and contrary to any simple form of 'interests' approach (see
chapter 5), it should be remembered that the pressures towards
a more cautious approach were to be found within the MRC.
Thus, whatever the locus of effective influence, discussions did
take place over whether the response should come from the MRC
or the ABRC. That is, there were negotiations about the breadth

of scientific interests, or at least the sponsorships, which should be represented within the emergent technological paradigm.

A cardinal feature of the British context within which these policy decisions took place constitutes an important difference from the USA and is not adequately dealt with in the cited studies. This concerns events following the escape of smallpox virus from the London School of Hygiene and Tropical Medicine in March 1973. There were two relevant consequences. First, a great deal of concern was aroused amongst the public, within the trade unions, especially the Association of Scientific, Technical and Managerial Staffs (ASTMS), and within the MRC, because scientists, and the system of control under which they operated, had failed. The details of the extraordinary series of lapses and failures were brought to light by a public enquiry into the affair, the Cox report (Cox, 1974). This had had a powerful effect on scientists working in laboratories dealing with infectious organisms, and on MRC officials concerned with laboratory safety. Procedures in many laboratories were quietly tightened up; microbiological safety had become a politically sensitive issue in Britain.

The second consequence was that a Working Party on the Laboratory Use of Dangerous Pathogens, under the chairmanship of Sir George Godber, was set up in November 1973. During the period of the Ashby and Williams Working Parties there were, as will become apparent in the following sections, several series of discussions about the relationship between the controls on dangerous pathogens and any possible controls on genetic manipulation research. In outline, there were two opposed positions. It could be argued that since it was responsible for public health issues, questions about the safety of genetic manipulation were formally appropriate to the Department of Health and Social Security (DHSS). On the other hand, a case could also be made for the DES to oversee the issue, since it involved research laboratories. As it happened, the Godber Working Party, which came within the ambit of the DHSS, received evidence on the implications of recombinant DNA techniques from Sir Hans Kornberg late in 1974. Its report, published in May 1975, recommended the establishment of a Dangerous Pathogens Advisory Group (DPAG) to control research in the area and also proposed that, DPAG 'should be so constituted as to meet the need for advice on control measures required in connection with work of the type considered by the Ashby Working Party' (Godber, 1975:5).

The first implications of the Godber approach were felt in May 1974, just two months before the publication of the 'Berg letter', when, because they felt that urgent action was required, the

Working Party presented an interim report to ministers. As an interested department, the DES had a representative, Dr Tyrrell, on the Working Party. In June a meeting was held at the MRC headquarters to consider the interim report, and to brief Tyrrell on the MRC response. The other participants included Dr Tony Vickers, an MRC official responsible for virus research, and a dozen or so scientists representing workers funded by the MRC. These included Brenner and Subak-Sharpe. At the time of the meeting it was known that the possible hazards of genetic manipulation had been brought to the attention of the Godber Working Party, and that it would shortly receive a report from Kornberg. There was more discussion at the meeting of the implications of genetic manipulation than of the interim report itself; in fact, this was the first formal discussion of the potential hazards of genetic manipulation within the MRC. With respect to the discussions concerning dangerous pathogens there was a consensus recognising the urgency of controls on a limited number of highly dangerous organisms, but there were reservations about the controls on less dangerous pathogens.

Accounts of the discussions relevant to the possible hazards of genetic manipulation techniques given in interviews have made it clear that the topic was not treated in isolation from other hazards of biological experimentation. Brenner, for example, gave a résumé of some of the more worrying possible outcomes of recent research developments, and argued strongly that these demanded more urgent consideration than the dangerous pathogens issues. These recent developments covered genetic manipulation, of course, but also included experiments such as those on cultures of human cells used in cancer research, the use of tumour viruses, and some cell fusion techniques. Brenner's feeling about the relative urgency seems to have commanded wide support from the participants, as apparently did deep concern over the potential risks implicit on the increasing entry into virology research of those without an adequate background in microbiological safety precautions. This echoed the concern expressed by Subak-Sharpe in his letter to the MRC about the impending 'Berg letter'.

At this early stage in the British debate over the recombinant DNA issue the possibility of hazard was being strongly emphasised to the MRC by knowledgeable and influential scientists. In contrast to later considerations of hazards, for instance in the Ashby report, genetic manipulation was discussed within the context of other biological hazards; and, in marked contrast to later opinions, there was no pressure to conceptually differentiate genetic manipulation from other techniques of biological experimentation, or to distance the issue from the interests of the Godber Working Party.[13]

Given the later views and actions of scientists this is surprising. The 'talking up' of the hazards of genetic manipulation was, of course, partly conditioned by the need to establish it as an issue worthy of attention. Even allowing for this effect, however, it remains clear that the likelihood of hazards was being taken very seriously indeed. But within a few months there had been a change of heart with regard to opinions about the interim report of the Working Party, and fears were being expressed about recombinant DNA techniques falling within the purview of Godber. It was then felt that the proposed controls on dangerous pathogens were particularly strict,[14] and that their application to genetic engineering would, in the words of one scientist, 'stop the field dead'. Possibly the full implications of the Godber approach had taken some time to sink in. For example, one of Brenner's arguments at the MRC meeting was for a code of practice to cover a range of the recent research developments, and it was implicit that such a code would require frequent updating as the science advanced. It may not have been realised that the thinking of the Godber committee on the institutional arrangements for the proposed DPAG would make such updating unlikely.

In general, the evidence of the MRC meeting indicates that it was scientists who were making the running in establishing genetic manipulation experiments as requiring some kind of control. There is no evidence of any attempt to play down the perceived risks, or of attempting to act in a narrow, self-interested, way; though it remains possible to see their actions as indicating that they had not yet perceived where their interests lay. The Cox and Godber reports are, therefore, an essential part of the context of the development of genetic manipulation policy in Britain – controls on dangerous pathogens, e.g. smallpox, had been seen to fail, and laboratory safety was a politically sensitive issue. The first discussions took place against the backdrop of the possibility of a role for Godber – or rather DPAG – in respect of genetic manipulation.

The Ashby Working Party

The Working Party on the Experimental Manipulation of the Genetic Composition of Micro-organisms was composed of thirteen members drawn from the ranks of the senior scientific establishment.[15] The ABRC had decided that, 'The membership of the Working Party should, as far as possible, not include those who were using the techniques and who might therefore be directly affected by its conclusions' (Ashby, 1975:iv). Commenting on this passage, Yoxen has pointed out:

> It is interesting to note that, for all their supposed independence in 1974, within three years six members of the committee were directing laboratories in which research on recombinant genetics was being carried out, and three of them have been directly involved in strategic planning by the British Agricultural Research Council (ARC) of programmes using recombinant DNA research. (Yoxen, 1979a:228)

Certainly, the general implication of the narrow circle from which the membership was drawn must be accepted; and this is an important factor in any evaluation of the Working Party's conclusions. But it must also be recognised that the scientific generality and significance of the new recombinant DNA techniques has been such that very few experimental research areas in biology could long have remained untouched by their application. The likely widespread use of the techniques was obvious at the time, though even the scientists were surprised by the speed with which they were introduced. None the less, the composition of the Ashby Working Party must be recognised as one of the steps in the evolution of the technological paradigm. Effectively, consideration of the recombinant DNA issue was still under a high degree of professional control.

The Working Party's terms of reference were: 'To assess the potential benefits and potential hazards of techniques which allow the experimental manipulation of the genetic composition of micro-organisms; and to report to the Advisory Board of the Research Councils' (Ashby, 1975:2).[16] The purely technical implication of the remit is emphasised early in the introduction to the report where the Working Party states:

> The problems we have been asked to assess have already received publicity in the press and on radio and television. They are causing interest and some concern beyond the boundaries of the scientific community. Indeed they are an example of a wider question, namely: how can the social values of the community at large be incorporated into decisions on science-policy? So it is in the public interest that our assessment should be intelligible to people unfamiliar with modern genetics. (Ashby, 1975:3)

Though not put strongly, the clear implication is that the report (which is admirably intelligible to the non-specialist reader) would not deal directly with the wider social issues. This implication is immediately reinforced: 'We were not . . . set up to make ethical judgements about the use of the techniques. Our business has been to assess potential benefits and potential hazards, after discussion with scientists who are familiar with this branch of biology' (Ashby, 1975:3). In subsequent sections the report details the potential benefits and hazards, noting that such an exercise must of necessity be a matter of speculation though resting on informed judgment (Ashby, 1975:6). In its conclusion the report states, 'We now have to declare our assessment of the potential benefits and practical hazards of using the techniques we have described. We reiterate our unanimous view that the potential benefits are likely to be great' (Ashby, 1975:12). And in the Recommendations, that 'subject to rigorous safeguards . . . these techniques should continue to be used because of the great benefits to which they may lead' (Ashby, 1975:15).

The rigorous safeguards were to be of two types, based on two different sorts of risk. Where the risks were of a known kind, precautions were to be modelled on existing forms of safeguards, of the kind used in the handling of infectious or pathogenic organisms. That is, existing forms of safety precautions well known in certain areas of biological research were to be introduced into genetic manipulation experimentation. This type of containment was later to become known in the field as 'physical containment'. Precautions are taken to reduce risks to laboratory workers (in the first instance) by physically closing possible routes of infection. The precautions are scaled to the danger from the organisms in question. They range from such practices as banning mouth pipetting, smoking, eating, and drinking in the laboratory (these restrictions are standard practice in laboratories dealing with infectious organisms) to, for example, the handling of samples inside sealed glove-boxes.

Physical containment was to be the primary safeguard (Ashby, 1975:11), but in addition to the use of accepted precautions in dealing with hazards of a conventional kind the Working Party proposed the use of 'genetical safeguards' or 'biological containment' measures appropriate to some of the particular conjectured hazards of genetic manipulation. One way would be to

> equip both the plasmids and the bacteria with genes which would disqualify them from surviving in the human gut at all. There are, for example, mutations which make the organism non-viable at normal body temperature and other mutations which restrict to special (suppressor) strains of

bacteria a plasmid's ability to replicate. There are other
mutations which make the organism dependent on chemicals
(an example offered to us was dependence on diaminopimelic
acid) which do not exist in the human gut. (Ashby, 1975:11)

In other words, the organisms with which scientists worked could
be genetically disabled or enfeebled to make it difficult or impossible for them to be a viable source of infection. This would be,
of course, to use standard genetic procedures to make genetic
manipulation techniques safer.

In addition to these standard and novel forms of physical and
biological containment a number of other precautions were suggested. In summary, they include the immunisation of laboratory
workers where relevant and possible, and epidemiological
(medical) monitoring of research workers. A biological safety
officer should be appointed in institutions where the experiments
were to be carried out, and there should be a well-publicised
advisory service available, perhaps run by one of the public health
laboratories. Those undertaking the experiments should have
some basic training in the handling of pathogens. In some cases
these suggestions are contained in a section of the Ashby report
which deals with 'some of the considerations which any body
drawing up a code [of practice] would wish to bear in mind'
(Ashby, 1975:13). In this area the report goes beyond any formal
interpretation of its terms of reference[17] by effectively laying down
the basic ground rules of a code of practice and providing a
number of exemplars. Obviously, this implies a particular answer
to the wider social and ethical questions in the debate. These
implications of the Ashby report are important not only in their
own right, but also because they represent the first publicly visible
signs of the general style of policy responses, and because of the
importance of the Ashby report as an input to the considerations
of the Asilomar conference.

Evidence presented to the Ashby Working Party

Twenty-eight scientists presented evidence to the Working Party,
some verbally. Part of the evidence is fairly technical and
concerned to develop arguments bearing on the existence or likelihood of possible risks from various types of experiment. All the
scientists were agreed that there would be considerable scientific
and medical benefits to be gained from the application of the new
techniques. The likely commercial and industrial benefits are less
frequently and less prominently mentioned, possibly because the
evidence stems from those with a background in academic science.

It is also noticeable, in contrast to more recent scientific opinion, but consistent with the MRC meeting on the interim Godber report, that certain categories of experiment are taken to be genuinely problematic in that the (possible) hazards are either (a) taken to be real, or (b) felt to require further evidence before any definite conclusion can be reached. Some categories of experiment were felt to be clearly non-hazardous. This is not to imply a solid consensus in the evidence, rather, that at this early stage an informal assessment of possible hazards was being made, and that in many cases the risks were being treated as 'real' possibilities.

Some scientists emphasised the radical uncertainty attached to the consideration of possible hazards. One statement from a group of five scientists working in Scotland noted, 'we just do not know what constitutes a "hazardous" DNA sequence, and we cannot recognise one' (Bishop *et al.*, 1974:9). They continued:

> We do not think it follows from this that work on mammalian DNA should not be done. The risk, like that in all medical research, is finite, but must be taken with the proper safeguards if either the hazards are to be recognised or the potential benefits achieved. We would certainly welcome a code of practice and an agreed protocol for this kind of experiment and believe that a full and open discussion of all results is essential. (Bishop *et al.*, 1974:9)

The statement is significant since the researchers in Glasgow and Edinburgh were recognised as being leaders in the field in Britain. Thus their 'welcoming' of a code of practice was likely to have been influential.

The evidence presented by Brenner indicates a slightly greater level of concern about the potential risks than that expressed by the scientists working in Scotland. For example he dismisses the frequently expressed defensive argument that the new recombinant DNA techniques were just a more convenient extension of experimental manipulations, and therefore no different in kind. He was also concerned that

> experiments could be tried by 'have a go' biologists who cannot control what they are doing with the present state of scientific knowledge. We would be justified in . . . fearing the worst one might accidentally produce microorganisms making proteins . . . which could produce undesired immunological reactions. (Brenner, 1974:7)

In view of Brenner's later, important, influence on developments in Britain and America, it is interesting to note that he made the following proposals in his evidence:

(1) Biological containment measures should be developed to reduce the survival of organisms outside the laboratory environment.

(2) A scale of hazards should be established and matched to a scale of acceptable protection measures. (Implicitly, currently used physical containment, plus biological containment.)

(3) A code of practice should be introduced (comprising e.g. (2)). This should be flexible, and the thresholds should be upgraded or downgraded as necessary as scientific knowledge increased.

(4) Administration and monitoring should be done by the research councils and other grant-giving bodies. The estimation of hazards should be part of the evaluation of research projects, along with such factors as the scientific quality of the proposed research. 'Highly hazardous experiments proposed by incompetent scientists could be excluded'. (Brenner, 1974:9)

The first point finds direct expression in the Ashby report (1975:16),[18] and points similar to the second and third are implicit in Ashby even though they are formally beyond the remit of the Working Party. The fourth point concerning the assessment of experiments by grant-giving agencies is echoed in the section of the Working Party's report that deals with the considerations that should be borne in mind by any body established to draw up a code of practice (Ashby, 1975:13). Elsewhere, and only in the particular context of potentially high risks, the report noted that there were relatively few laboratories in Britain where such experiments could be performed, and 'anyone contemplating high risk experiments should first of all secure the agreement from his peers that the experiments really need to be done (by reason of the scientific or social benefits which might follow from them)' (Ashby, 1975:11). Such points, or close relatives of them, were to appear throughout the course of development and implementation of British policy on genetic manipulation. In view of this it is relevant to note that the assessment of the need to perform certain classes of experiment is here placed firmly within the context of professional assessment and control.

The extent to which the evidence of Brenner (and others) influenced the final recommendations of the Ashby Working Party must remain a matter of some uncertainty. Similar proposals to those just noted were expressed by Brenner at the Asilomar conference, and his influence there has been well documented (Rogers, 1975).

In summary, then, the evidence available on the workings of the Ashby committee indicates that, in effect, the scientists were using their scientific and perhaps political knowledge to fill in the details of the technological paradigm. But there are indications that this cannot be viewed in terms of a simple self-interested approach. A good many scientists had accepted the likelihood of possible hazards and the existence of serious concern, and were working within the terms of their particular technical competence to produce an accommodation.

In short, the issues surrounding genetic manipulation emerge from Ashby not as problems, but as puzzles in need of a technical solution, to borrow Kuhn's (1970) well-known distinction. Within this 'paradigm' of internal professional control scientists were to scale hazards and precautions, to construct appropriate measures of physical and biological containment, and to flesh out the details of the code of practice.

Thus the Ashby report emphasised the potential benefits of genetic manipulation research 'subject to rigorous safeguards'. These include the training of workers in the handling of pathogens; the establishment of safety officers in laboratories, and a central advisory service; epidemiological monitoring; special precautions for large-scale experiments; physical and biological containment measures; and research on the survivability of *E. coli*. In addition, a code of practice was thought to be necessary and the Working Party devoted eight paragraphs to detailing factors which were thought should be considered in drawing up such a code.

Developments between the Ashby and Williams Working Parties

The publication of the Ashby report was quickly followed by the international conference at Asilomar in February 1975. Broadly, the final report of the Asilomar conference, which had Brenner as one of its five authors, agreed with Ashby that most experiments in the area should be allowed to go ahead, subject to suitable safety precautions. In a number of respects the recommendations of Asilomar can be seen as later developments of themes that are implicit in the Ashby report. The conference report emphasises that containment measures must be made to match the degree of estimated risk. It also makes a now familiar point:

the ways in which potential biohazards and different levels of containment are matched may vary from time to time, particularly as the containment technology is improved. The means for assessing and balancing risks with appropriate levels of containment will need to be re-examined from time to time. (Berg *et al.*, 1975:442)

Thus, in general terms the Asilomar conference supported and extended the view that existing modes of physical containment normally used in research on infectious organisms could be applied to the recombinant DNA issue. The conference differed from the Ashby report in placing much greater emphasis on the potential for biological containment through the development of disabled strains of *E. coli*.

Following the spell of activity during the winter of 1974–5 there was something of a hiatus in Britain. Whilst the American Recombinant DNA Advisory Committee went ahead with developing specific and detailed guidelines for various categories of experiment, visible official action in Britain was nil. There had been a plan for the Royal Society to hold a follow-up scientific conference to promote action on the Ashby recommendations, or at least to discuss them, but nothing came of it. In July 1975 a conference was convened at Oxford, under the auspices of the MRC's Cell Board. The meeting, which was attended by some 120 scientists, was organised by Brenner and Professor Walter Bodmer, then an Oxford geneticist. In part, the conference was intended as a forum within which recent events could be discussed, and current scientific opinion gathered.[19] But more importantly its role was conceived of as a forum within which scientific opinion could be mobilised. 'You see, Ashby had sat, and then nothing happened . . . So we set up the Oxford meeting in order to put pressure . . . to do something' (Brenner, 1981). Four scientific sessions were organised which dealt with various aspects of recombinant DNA techniques. A separate session was devoted to a discussion of the issues of possible hazards, of safety precautions, and of the possible institutional arrangements that might be introduced to administer or oversee any code of practice. In this session it became clear that many scientists were becoming impatient. Many more laboratories were reaching the stage where they would like to undertake genetic manipulation experiments, but there was as yet no code of practice, or any institution capable of producing one. There was a feeling that if the Ashby recommendation of a central advisory service – which would give scientists some idea of what experiments were allowable – were not soon implemented, a few scientists would begin to ignore the requests of the 'Berg letter'. Indeed, the journal *Nature*, which had greeted the Ashby report as the 'Amber light' for genetic manipulation, ran a leading article on the Oxford conference under the title of 'Forever amber on manipulating DNA molecules?', and reported that:

> It was wrily conceded that more impatient scientists were already under way and although there was no evidence that they were conducting experiments in an irresponsible way, their action was causing considerable frustration amongst those waiting for more formal guidance on safety. (Anon, 1975:155)

In addition to the successful mobilisation and display of this general pressure for action, support was also generated for a number of more specific objectives. These were a series of preferences about the form and relationships of what became called the Central Advisory Group (CAG), the erstwhile central advisory service of the Ashby report. Broadly, support from the meeting was forthcoming for the view that the CAG should function much as anticipated by Ashby, but a number of riders were added including

(a) That the Central Advisory Group should report to the DES.
(b) That, since the functions of the CAG were to deal with unknown, hypothetical, hazards, it would not be appropriate for it to be a part of the Dangerous Pathogens Advisory Group (DPAG) recommended by the Godber Working Party, whose role would be to deal with known, real, hazards.
(c) That an early objective of the CAG should be to define its sphere of operations in relation to genetic manipulation techniques, leaving a review of other safety issues on biological experimentation to the MRC itself.

In the light of the earlier discussions about the development of British policy on genetic manipulation it is clear that these objectives did not emerge spontaneously from the meeting. As far as is known they were worked out by members of the senior scientific establishment, especially Brenner, in conjunction with the MRC officials. There is no doubt, though, that a broad consensus for the objectives was generated at the conference. The objectives can be seen as a fleshing out of the technological paradigm, or at least as part of an attempt to ensure professional control over the code of practice. For example, objective (a), that the CAG should come under the DES rather than the DHSS, has the ostensible (and perfectly defensible) rationale that the former is more used to dealing with laboratory matters. A necessary consequence of this would be that the genetic manipulation issue would be retained within an arena more permeable to influence by scientists. Many scientists at the conference were no doubt pleased to hear the DHSS representative announce that his department would not be making a bid for the CAG, though it should be noted that this statement need not necessarily mean that the CAG would not eventually fall to that department, or that genetic manipulation techniques would not be dealt with by the Dangerous Pathogens Advisory Group. Several participants commented that after the conference they still felt the need to maintain the pressure to avoid DPAG.

The three objectives are interlinked. Objective (b), the creation of a CAG independent of DPAG, is obviously complementary to (a). The argument for these linked objectives tended to be presented thus: since the hazards involved in genetic manipulation and dangerous pathogens research were of different kinds (i.e. unknown and known), and because of the difficulty in extrapolating to hypothetical hazards from real ones, then it followed from this that different institutional arrangements were called for. In fact, this seems to be almost a reversal of the actual reasoning. It would be more accurate to say that because of the perceived need to separate the institutional arrangements it follows that the hazards issues must be treated as being of different kinds. A related point can also be made with respect to objective (c) concerning the relationships with other forms of hazard in biological experiments, and the need to differentiate their consideration. In both cases the 'scientific' aspects of the relationships between the hazards were flexible; questions of strategy were decisive.

It is clear that at the time of the MRC meeting to discuss the implications of the interim Godber report in 1974 biological hazards were being treated as if they were all of a piece, and there was no discernible pressure to distance genetic manipulation from dangerous pathogens. This opinion changed within a few months. By the time of the Oxford meeting scientists would have had the chance to read the final report, which was published in May 1975, and which recommended that DPAG should be so constituted as to be able to provide the central advisory service proposed in the Ashby report. Such an eventuality was now regarded with extreme concern by many senior scientists, and their view seems to have been shared by MRC officials. Thus there were 'good reasons' for the reversal of logic in the argument for objective (b). The institutional separations proposed in objectives (a) and (c) are, in fact, part and parcel of strategic decisions that were more or less explicitly taken at the Oxford conference, and which involved scientific decisions about the way in which risks were defined as similar or different.

The question of whether or not risks (or any other classes of objects or categories) are *the same* or *different* necessarily begs the question of all similarity relationships – 'Similar with respect to what?' There is an indefinitely large number of similarity or difference relationships; and the reason why one set rather than another are selected as important in any particular context will depend significantly on the purposes at hand.[20] From the point of view of the scientific predictability of hazards, the distinction between known and hypothetical hazards can be argued to be absolute and vital. In a similar vein, the assessment of the safety precautions required for dangerous pathogens and genetic manipulation research may well require different forms of scientific and medical expertise. On the other hand, it could be argued that there is a similarity in that both are risks involved in biological experimentation, that the distinction between 'known' and 'unknown' hazards is relative, and that both require a probabilistic analysis. During the first discussions within the MRC a much wider set of biological risks was considered, and it might seem only reasonable that some rough assessment of the relative scale of biological hazards would have been made, especially given the earlier opinion that the development of a code of practice relevant to genetic manipulation techniques was of greater urgency than that for dangerous pathogens.

Such a comparative note was raised at the conference by Dr Robin Weiss of the Imperial Cancer Research Fund. His argument, in a paper published in *Nature* (Weiss, 1975), sets the possible hazards of genetic manipulation in a context which demands a consideration of the relative risks involved. Essentially Weiss points to the possible hazards associated with some forms of routine, accepted, and largely uncontrolled areas of biological research (not genetic manipulation). For many scientists the immediate implication was that by comparison genetic manipulation was not clearly so worrisome after all. There were, however, two divergent views as to the wider consequences of articulating such a perspective.

If it were publicised that the genetic manipulation techniques were probably less potentially hazardous than other, accepted, kinds of biological experiments this might lead to a reduction of pressures for strict controls. But, of course, the opposite could apply. Public concern might become focused on the wider question of the possible hazards of biological research, risking the possibility of codes of practice and controls on a very wide range of experimentation. This might well have covered such techniques as cell fusion and tumour virus research as were discussed at the MRC meeting on the interim Godber report. Given the scientists' perceptions of the level of precautions they expected to be introduced, this would be a disaster for areas of research which were not then under threat.

The participants at the Oxford conference were faced with a dilemma. There can be little doubt that the scientifically respectable courses of action was to publicly air the broader question of the comparability of the possible hazards of the various kinds of biological research. But, politically, this would have been a high risk course of action which could have been successful in ameliorating the likely stringency of control only if scientists were able to create public confidence in the view that, in general, existing practices in biological research represented adequate safeguards. Given the background of the London smallpox outbreak, and the initial reaction to the 'Berg letter', this seemed unlikely. Evidence from interviews shows that there was fairly strong resistance to the high risk alternative. In particular it was identified as coming from scientists not involved in genetic manipulation whose own research might be affected. There was a certain amount of informal pressure on those who agreed with Weiss's view not to 'rock the boat'. As already noted, Weiss's views were published in *Nature*, but did not produce any reaction in the correspondence columns of that journal. Effectively, the wider questions about the possible hazards of biological research were contained, either by being rejected or ignored, and then set aside. The conference implicitly followed the lower risk alternative, and scientific support was mobilised behind the view that any hazards associated with genetic manipulation were different and should be dealt with separately.

This episode is symptomatic of the scientists' position at the time. On such scientific matters there was a limited amount of room for manoeuvre open to them, particularly the 'definition of the situation' with regard to the hazards; but their undoubted scientific authority was not sufficient for them to feel able to adopt the high risk alternative. The agreed scientific view, or perhaps the agreed scientific platform, adopted was the result of taking into account social and political factors. In itself this is a clear indication of the scientists' perceptions of the relative strengths of scientific and public opinion over the question of hazards.

Before leaving this particular topic, a further pressure towards the low risk alternative, or at least not rocking the boat, should be noted. This pressure came from the MRC officials, who were aware that the government response had already been made, and that the Williams Working Party would shortly be announced. This occurred on 6 August 1975. It was thought far better for scientists to continue to play a low-key role as their feelings at the meeting would in any case be taken into account by the Williams Working Party. Six of its ten members, including the chairman, were participants at the Oxford conference.

The opinions expressed and decisions taken at the Oxford conference partly reflect a slight lessening of fears among scientists about the severity of the hazards that might be associated with recombinant DNA techniques. This is not to say that the early concept had evaporated, but to mark the scientists' perception that there were many minimal or low risk experiments as safe as, or safer than, other biological experiments which were being performed daily. Given these were accepted as safe, there seemed no reason for not going ahead, particularly as the Americans were busy constructing guidelines which would soon allow many such experiments to be performed with only limited safety precautions.

At the same time there was also a desire for some system or set of principles with which to categorise more potentially hazardous experiments. Some of these held great scientific interest, even though the scientists expected the containment levels to be severe, and the work therefore difficult to perform. As evidence of this, there were discussions at the Oxford conference initiated by Brenner about the sharing of the limited number of high containment facilities that were available, including the possible use of laboratories at Porton Down, the former biological warfare establishment. Following these and related discussions, the conference felt the need to propose the establishment of a body to represent the interests of the scientists, and to aid the sharing of high containment facilities. This body, to be run by the MRC, later became known as GMUG, the Genetic Manipulation Users' Group. These ideas are indicative of just how serious some categories of hazard, and hence how rigorous the expected controls, were then taken to be.

The decision to adopt a low risk strategy was perhaps the only politically sensible course. So far the British developments had been handled in a low key manner without what many scientists saw as the histrionics of the American debate. Opening up the more general question of the safety of biological experiments would run the risk of its closer association by the public with the consequences of the smallpox outbreak. Were this to happen the arena of discussion and hence that of effective decision making might well be broadened, with a corresponding reduction in the ability of scientists to influence the outcome.

The effect of the Ashby report was that the issues surrounding recombinant DNA research had become defined as technical safety matters demanding professional attention within the scientific community. The Oxford conference did nothing to disturb this; it reinforced it whilst feeding in a set of more specific objectives.

The emergence of the trade unions strategy

The trade unions, especially the Association of Scientific, Technical and Managerial Staffs (ASTMS), appear to have had a considerable influence on the development of, and later on the operation of, genetic manipulation regulations in Britain. In part this was due to their ability to marshal expert scientific opinion from within their membership, and to their access to the decision making arena (Wright, 1978). Also important in this respect was the legislative framework of the 1974 Health and Safety at Work Act (HASAWA), a topic which will be discussed shortly.

By March 1976, when the Williams Working Party was considering the details of a code of practice for genetic manipulation experiments, and the appropriate institutional arrangements for implementation, the ASTMS position was centred around five points. First, genetic manipulation was, in trade union terms, a progressive technology in that, in addition to the expected scientific and industrial benefits, it was also clear that there would be important medical and diagnostic advances. Thus, subject to appropriate safety precautions, the research should be allowed to go ahead, and should be well funded by the research councils. Second, there should be enforceable regulations under the HASAWA. Third, there should be a central advisory service which would also maintain a compulsory register of all genetic manipulation experiments. Fourth, the central advisory service should have a strong trade union representation; and fifth, before any experiment could be submitted to the advisory service it would have to be agreed by the trade union safety representatives in the laboratory.[21] With the possible exception of the first point these views were presented to the Williams Working Party as a TUC (Trades Union Congress) approach by a delegation which included representatives of ASTMS and NALGO (National and Local Government Officers' Association). The importance of these views can be recognised by the fact that they were largely incorporated into official policy. Only a limited amount of detail is available on how the policy was developed and refers mainly to ASTMS, which seems to have played the dominant role.[22] The somewhat fragmentary data are of interest because they indicate the early, exploratory moves in the process of policy formation. The apparent solidity of the ASTMS position at the time of Williams was presaged by a potential divergence of approach which, although successfully contained, illustrates a recurrent tension between an arms-length trade union approach to health and safety issues, and the specific professional interests of the scientist trade union members.

At the time of the 'Berg letter' much of the British experience and practical knowledge about genetic manipulation techniques was concentrated in Scotland. In particular, the Beatson Institute for Cancer Research in Glasgow; the MRC Virology Unit, the Virology Department and the Biochemistry Department at the University of Glasgow; and the MRC Mammalian Genome Unit and the Department of Molecular Biology at the University of Edinburgh were the important centres. The scientists at these laboratories had an active interest in the eventual outcome of decisions about the future of genetic manipulation. In September 1974 five of them had presented evidence to the Ashby Working Party stressing the likely potential benefits of the new techniques, and welcoming a code of practice to deal with any possible hazards (Bishop *et al.*, 1974). It also happened that each of the centres had a strong ASTMS organisation, and in 1975 there was a meeting of some twenty ASTMS representatives from these laboratories to discuss the issues surrounding genetic manipulation, and the recently published Ashby report.

Broadly, the meeting supported the conclusions of the Ashby Working Party. On the question of a mechanism through which its report's recommendations could be implemented, the ASTMS members were in favour of the establishment of an advisory committee under the auspices of a research council (presumably the MRC) which could offer advice to research groups, and ensure that the code of practice was followed. This, of course, is perfectly compatible with one of the important policy implications of the Ashby report: that decision making would be kept within scientific circles. The ASTMS members were, however, of the view that the advisory committee should have trade union representation, a move which would broaden its constituency, but which would still not necessarily mean that non-scientific or outside interests would be represented.

These scientists were convinced that many genetic manipulation experiments could be safely performed, provided that appropriate safety precautions were observed. Thus this first position was a reflection, on the one hand, of both their expert judgment about the nature and magnitude of any possible risk, and their scientific interests in wanting to proceed with experiments, and, on the other hand, their trade union perspective that the best form of safety advice would be generated by a body which contained trade union representation.[23]

It is worth emphasising that there was no mention – at least in participants' recollections – of the question of any statutory basis for, or involvement of the Health and Safety Executive (HSE) in the administration of, a code of practice. Indeed, neither seems to have been raised as a possibility at this point. The tenor of this early position is reinforced in a set of briefing notes prepared for ASTMS in January 1976 by Professor Bob Williamson, then a senior scientist at the Beatson Institute, and a co-convenor of the meeting just mentioned. Williamson, a long-standing and committed ASTMS member, has played a significant part in the development of ASTMS policy, and has often been cited in interviews as one of the influential members of GMAG. By the time that the briefing notes were prepared, full-time ASTMS officials had already had some exploratory discussions with the HSE, but it is doubtful that this information was generally available to members, and Williamson's notes contain no indication that the HSE might have a part to play in the control of genetic manipulation in Britain.

Williamson, writing in a personal capacity, made it clear that his briefing notes represented a proposal for an immediate policy for ASTMS, and the TUC. He noted that the meeting of ASTMS members based in Scotland had effectively endorsed the general implications of the Ashby Working Party (and the implications of developments in the USA) that 'the new [genetic manipulation] techniques were seen as being part of a general and continuous scientific advance in molecular biology, which neither should nor could be interrupted' (Williamson, 1976:1). Briefly, Williamson's recommendations for an immediate policy were these:[24]

(1) That the Advisory Board to the Research Councils should, perhaps through the Williams committee, allow some work to go ahead immediately. Since no UK guidelines had been established it would be as well to adopt those of the NIH.
(2) There should be trade union representation on any

advisory committee recommended by the Williams Working Party.

(3) In the case of the very few experiments where there might be real safety problems a representative laboratory safety committee should determine whether the experiment could be done at all.

(4) There should be centrally run training courses.

(5) Centrally run 'safe facilities' (containment laboratories) should be made available to researchers.

(6) Research councils should be provided with substantial funds to promote genetic manipulation research.

Williamson wrote as a committed ASTMS member, and as a scientist. He was concerned that although work had started up again in other countries, 'Two years after the moratorium . . . and a year after the Ashby committee . . . there is still no code of practice and safety, no consultation mechanisms, no decision-making process for research in this field in Great Britain' (Williamson, 1976:2). He noted with alarm that Britain was falling behind in the research, and that 'individual scientists are conniving either with scientists abroad, or in their own laboratories, [to] circumvent the restrictions, which is more likely to entail risk than would properly discussed experiments' (Williamson, 1976:2). The briefing notes were part of the process of the formation of ASTMS (and TUC) policy. Naturally this involved processes of negotiation, many of which remain hidden, but a glimpse is afforded in an expansion of point 6. Williamson noted there was a risk that

> the debate over safety will be used as an excuse in the present financial climate to delay work and expenditure . . . it is quite important, in my view, to ensure that trade union representatives should not allow themselves to be used as excuses for financial deferments which they would under most circumstances oppose. (Williamson, 1976:3)

Clearly Williamson was concerned to head off any potential move towards an overly restrictive approach to the progress of his science, either through decisions about funding or safety. Precisely how strong any such pressures might have been within trade union circles is partly a matter of conjecture, especially since participants' recollections may be coloured by the currently accepted view that the original fears about genetic manipulation were overstated. All the same, it does seem that within the conspectus of opinion there were those who tended towards a more cautious approach and that such opinions were somewhat more likely to be found amongst full-time union officials.

Circumstantial evidence for this can be found in the fact that ASTMS has a reputation for taking a strong line on health and safety issues. Wright has cited an undated ASTMS document from the 1970s that 'Health and Safety must never be sold for short term benefits. No agreements on danger money, risk payments, etc., are acceptable' (Wright, 1978:1427). Obviously, ASTMS would be vitally concerned with issues affecting the health and safety of laboratory workers and, especially after the revelations of the Cox report on the 1973 smallpox outbreak from the London School of Hygiene and Tropical Medicine, was well prepared to use its powerful political influence.

In March 1975 ASTMS representatives had met Reg Prentice, the Labour Secretary of State for Education and Science, to discuss the Ashby report. In June 1975 there was a meeting between ASTMS and HSE at which both the Ashby and Godber (Dangerous Pathogens) reports were discussed. By now the full-time officials representing ASTMS were clear that their union should be involved in the implementation of the recommendations of both reports. ASTMS envisaged a set of controls administered by a single body (perhaps in line with the Godber report) covering both dangerous pathogens and genetic manipulation. Also clear at this stage was the ASTMS officers' view that some legally enforceable mechanism of control was urgently required, and that the registration of laboratories engaged in genetic manipulation should be compulsory.

For the trade unionists the possible hazards of the new techniques were part of a larger context of concern for the health and safety of laboratory workers, particularly those employed in health service laboratories. ASTMS had developed an antagonistic attitude towards the manner in which the DHSS dealt with health and safety issues in health service laboratories. Existing codes which were available were often not properly implemented, as witnessed by the comments in the Godber report. The union was keen to change this. They intended to appoint a safety officer to co-ordinate their activities, and they were hopeful that the HSE would make its presence felt in the area. Other things being equal, genetic manipulation would become one aspect of this general policy, which would be pursued vigorously.

Thus the ASTMS policy position at this juncture was centred on a vigorous, but standard health and safety approach.[25] In distinction to Williamson's later briefing notes, and perhaps an expression of the tendency which prompted them, there is no mention of any special status for genetic manipulation dependent on the hypothetical nature of the possible risks, nor of any sense of urgency consequent on the need to allow experimentation to go ahead. Where urgency is expressed it is in the context of reducing the risks in health service laboratories. The Oxford conference had occurred between the ASTMS/HSE meeting and the Williamson briefing notes.

These sets of opinions were not necessarily incompatible; indeed, they were later integrated into a coherent policy. Further, it should be remembered that Williamson was also concerned with health and safety issues, and that ASTMS has a tradition of bolstering its arguments through appeals to the economic advantages of supporting appropriate forms of technical innovation. The point is rather that these were potentially divergent positions that were accommodated through negotiations within ASTMS. The tension between a more traditional trade union approach and the immediate professional interests of scientists was not, however, permanently solved, and it surfaced again during the first year of GMAG, when the debate was much more polarised.

A cardinal piece of context relevant to both the trade union strategy and the development and operation of genetic manipulation regulations was the passing into law of the Health and Safety at Work Act (HASAWA) in July 1974. The Act was part of a package of Bills with which the Labour Government replaced the Conservative Industrial Relations legislation. Broad in scope, the aim of the HASAWA was the eventual consolidation of various existing pieces of health and safety legislation, and the gradual extension of similar provisions to increasing numbers of workers. The HASAWA established a Health and Safety Commission (HSC) to oversee the work of the HSE, which incorporated such existing bodies as the Factory Inspectorate and the Employment Medical Advisory Service.

The Act laid a duty on employers to take all reasonable steps to protect the health and safety of their employees, and members of the public, from workplace hazards. The Secretary of State for Employment could, on the recommendation or advice of the HSE (via the HSC), approve codes of practice for various occupations, industries or processes. Such codes of practice would not in themselves be legally binding, but in any case brought before the Courts, failure to comply with the provisions of a relevant code would, *prima facie*, be held to be a failure to have taken 'all reasonable steps' to ensure safety.

Essentially the HASAWA was enabling legislation. It gave wide powers to the Secretary of State to act without any further recourse to Parliament. More importantly for the present discussion, it institutionalised the role of the trade unions in the process of developing codes of practice since, for example, they were to be included in the extensive consultation procedures of the HSE, and one-third of the HSC were trade unionists. Other provisions of the Act included the right of employees to elect workplace safety representatives (paid by the employer), the right of workers to a written statement of their employer's health and safety policy, and the establishment of a formal requirement on employers to consult workplace safety representatives and safety committees in the creation and operation of health and safety procedures. The HSE itself was able to make arrangements for research, for the dissemination of information, and for training on relevant health and safety issues.[26]

Thus the HASAWA was potentially available as an organisational framework for dealing with some of the safety issues raised by the fears about genetic manipulation. Precisely when this potential was first realised or seriously considered is difficult to establish – and, no doubt, it was not a single event. An early recorded comment is in a letter to *Nature* published in August 1974 (during the deliberations of the Ashby Working Party) by Tannahill of the Employment Medical Advisory Service, then a part of the Department of Employment. In response to an article by Ford (1974) calling for legislation to cover experiments with potentially dangerous biological agents, including genetic manipulation, Tannahill pointed out that 'workers and other persons' could be protected 'from risks to health in connection with the use, handling, storage and transport of articles and substances' under the HASAWA without the need for any further legislation (Tannahill, 1974:618). Given the formal availability of this information it is perhaps surprising that, in the early considerations of genetic manipulation by the trade unions, little account seems to have been taken of the potential of the HASAWA. This is despite the fact that the new Act was apparently widely discussed in laboratories. Seemingly it took some time before the practical implications of the HASAWA for genetic manipulation were realised.

However, the fact that laboratory workers were, for the first time, formally covered by the legislative framework of the HASAWA had to be translated into a set of institutional arrangements.[27] At the time of the ASTMS/HSE meeting such matters were under initial consideration by HSE officials, who also tended to the view that there should be a single system of control for the hazards covered in the Ashby and Godber reports. Establishing such mechanisms would take time, given the necessary, extensive, consultation procedures of the HSE. With the Williams Working Party on the horizon (it was formally announced on 6 August 1975, but its existence was already known), there would have been little reason for an immediate HSE proposal, especially given the technical knowledge which this would require. It seems then that the HSE interest in the area, and hence that of the trade unions, was well recognised within the relevant government departments. However, the organisational arrangements for dealing with genetic manipulation research were formally part of the terms of reference of the Williams Working Party.

3
Establishing the Genetic Manipulation Advisory Group

The Williams Working Party on the Practice of Genetic Manipulation was formally announced in the House of Commons by the Secretary of State for Education and Science, Fred Mulley, on 6 August 1975. The report, published in August 1976, contained recommendations for the establishment of a Genetic Manipulation Advisory Group (GMAG), though not its precise composition, together with a system of local safety committees and safety officers, a detailed code of practice for genetic manipulation experiments and a system for their categorisation. Other recommendations covered such matters as training, and the medical and epidemiological monitoring of laboratory workers.

The first part of this chapter concentrates on the major area of policy decisions – those related to the institutional arrangements within which advice on, and control of, experimentation should take place. The code of practice for experiments is contained in an extended and detailed appendix to the Williams report. This, and issues such as training and medical monitoring, will be dealt with only in passing, and where they are germane to the more consequential policy decisions. The second part concentrates on the discussions of the principles underlying the basis for assessing risks at the experimental level, while the third deals with the terms of reference and composition of GMAG.

The Williams Working Party

The Williams committee was composed of ten members including the chairman, Sir Robert Williams. There were also, as is usual on such committees, assessors from interested government departments – in this case one from each of the Department of Education and Science, Department of Health and Social Security, the Ministry of Agriculture, Fisheries and Food, and the Scottish Home and Health Department. Two secretaries to the Working Party were provided by the Department of Education and Science. The members were almost exclusively drawn from the ranks of the senior scientific establishment; more specifically, seven of them were the Directors of important research institutes. Indeed, the only person not directly involved in performing or directing biomedical research was Dr Ron Owen of the Health and Safety Executive.[1] Of the ten members, six had been present at the Medical Research Council's meeting at Oxford in June 1975, and four of them were also members of the Godber Working Party on the Laboratory Use of Dangerous Pathogens. Williams had been a member of the Ashby Working Party. It is also worthy of note that seven of the members of the Williams committee (including Owen) were later to become members of GMAG, or co-opted members of its subcommittees.

The experience and expertise represented on the Working Party was fairly broad at the technical level. The chairman was Director of the Public Health Laboratory Service in London. Other members came from laboratories where research involved, or was concentrated on, such areas as molecular biology, plant genetics, dangerous pathogens, cancer, and animal viruses. The majority had experience of the application of one or more of the mainly voluntary codes of practice which already covered certain areas of biomedical research. At this early stage perhaps a third of the scientists had knowledge and expertise of the recombinant DNA techniques.

The formal terms of reference of the Working Party were:

In the light of the reports of the Advisory Board for the Research Councils Working Party on the potential benefits and potential hazards associated with the genetic manipulation of micro-organisms and of the Working Party on the Laboratory Use of Dangerous Pathogens (a) to draft a central code of practice and to make recommendations for the establishment of a central advisory service for laboratories using the techniques available for such genetic manipulation, and for the provision of necessary training facilities; (b) to consider the practical aspects of applying in appropriate cases the controls advocated by the Working Party on the Laboratory Use of Dangerous Pathogens. (Williams, 1976:3)

Clearly this placed a potential conflict on the agenda, since the Ashby report had recommended a voluntary code of practice on the grounds that a statutory code would be unreasonably difficult to administer, whereas the Godber Working Party had proposed that the Dangerous Pathogens Advisory Group (DPAG) should be so constituted as to be able to advise on, and control, genetic manipulation techniques. However, beyond these terms of reference, less formal influences and constraints operated. As noted in chapter 2, the Health and Safety Executive (HSE) and ASTMS/TUC view favoured a single system of control for genetic manipulation and dangerous pathogens, whilst the Oxford conference had rejected Godber's recommendations. The Williams Working Party was under pressure from scientists to produce its report as rapidly as possible. The development of guidelines in the United States added to the urgency since it would be awkward, to say the least, if Britain lagged too far behind. In addition it was important that the British guidelines did not produce radically different assessments of risk, or categorisations of experiments.

A second set of constraints revolved around the extent and nature of the representation of the interests of employees in the operation of the advisory service or control body eventually to be recommended. The Secretary of State had agreed that the trade unions, and specifically ASTMS, would, at the very least, be involved in consultations about the form of the advisory or control system, and most probably in its operation. The Health and Safety at Work Act (HASAWA) was central to trade union ambitions for influence and involvement in the area. The Act was also important as a more direct constraint on the possibilities of decision making open to the Williams Working Party, since any recommendations would have to take account of the interests of the HSE, and consequently the role of the trade unions. Third, public fears and concern, at least as expressed in Parliament, would have to be allayed. The political reality of the situation required that an acceptable, safe set of procedures should operate, and should be seen to operate.

None of these constraints and influences were strong enough to force the Working Party into any particular course of action; rather they were items that were fairly immediately recognised as being on the informal agenda. The final introductory comment on the Working Party has to do with a patterning of its work which, in part, was not visible to the members. The press release announcing the establishment of the Working Party noted that 'The Government accepts that it has a responsibility to ensure that *authoritative* advice and guidance are available' (MRC, 1975:1, emphasis added). For scientists an important aspect of the authoritativeness of the advice and guidance offered would be the scientific legitimacy or reasonableness of the details of the code of practice, the assessment of possible risks, and the appropriateness of the precautions.[2] The recommendations of the Working Party would necessarily be evaluated by scientists in the light of their technical knowledge. That is not to say that there was any significant expectation of a watertight scientific argument for safety precautions – indeed, there was a high degree of expectation that political or social factors would necessarily be involved. Scientists did, however, expect a certain scientific rationale for the advice they would be expected to follow.

For non-scientists a second meaning of authoritative was probably more important. For those unable to follow in detail, or assess individual classifications of risk, or the appropriateness of the code of practice, judgment was likely to be focused on the characteristics of the formal mechanisms through which decisions about experiments would be reached, and on the likelihood that the advice would be followed. In short, outsiders would pay as much attention to the bureaucratic characteristics of authoritative advice as to the scientific. A degree of bureaucratic expertise was therefore required in order to produce a system which could offer authoritative advice. Necessarily, then, the deliberations of the Williams Working Party, as those of other bodies in the area, would take place within the context of an interaction between scientific, administrative, organisational and political factors. As noted in chapter 1, there is a wide measure of agreement that social and scientific factors have both been involved in the creation of guidelines and codes of practice for genetic manipulation. By common scientific consent, there were no 'facts' available on any 'real' or 'proven' hazard. Thus scientific beliefs or courses of action were not simply or rigidly determined by 'the facts of the matter' but were created to a purpose, as in other walks of life. In discussing the Working Party a perspective will be drawn upon from the British 'relativist' tradition in the sociology of science which shows how scientific knowledge is constructed through the flexible deployment of available cultural resources (Mulkay, 1979).

The Central Advisory Service

It has often been remarked that GMAG, which was developed from the earlier idea of a central advisory service, was a novel response to a novel situation. Certainly, there is no precedent for the moratorium called for in the 'Berg letter', and there has been no committee on science policy in Britain where representatives of the interests of management, employees and the public has been so well institutionalised. However, it appears that the novel aspects of GMAG owe little to the initiatives developed within the Williams Working Party itself. Furthermore, a close look at the development of the thinking of the committee reveals the extent to which other aspects of the advisory service recommended were modelled on, or constructed from, other organisations concerned with health and safety in biomedical research.

Before the first meeting of the committee Sir Robert Williams circulated a paper to members proposing that potential hazards might be structured or scaled in terms of degrees of containment. Four levels of containment were suggested (A, B1, B2 and C), the first three of which were closely modelled on the report of the Godber Working Party, of which Williams had been a member.[3] The containment levels were also loosely related to the levels of risk discussed at the Asilomar conference. The question of direct comparison is difficult, but a rough correspondence would be: Category A = Asilomar high risk; B1 = medium risk; B2 = low risk. The lowest of the levels of containment suggested by Williams, category C, would correspond roughly to the normal practices obtaining in microbiology laboratories, where, for example, experiments are performed in open laboratories without any destruction of the organism when the experiment is complete.

Category B2 would approximate the procedures routinely used in laboratories dealing with potentially infectious organisms, for example, in public health laboratories. Such work is normally performed on open laboratory benches, but mouth pipetting, smoking, eating and drinking in the laboratory are prohibited. Equipment and samples are normally sterilised after use. Category B1 would involve a more rigorous set of containment procedures and practices. Experiments are performed inside glass safety cabinets which are maintained at an air pressure lower than that of the laboratory. This helps to prevent any aerosols generated during an experiment, and which could contain bacteria or viruses, from escaping into the laboratory. Sterilisation procedures are extensive. Such conditions would normally be appropriate for dealing with more robust, pathogenic organisms such as tuberculosis bacillus, for example.

The most stringent level of containment (category A) was that advocated by the Godber Working Party for experiments on dangerous pathogens such as smallpox virus. Protective clothing and masks are worn, and experiments performed within glove boxes sealed off from the operator and from the laboratory. In turn, such laboratories are themselves usually sealed off from corridors with entry and exit through an air-lock. Air pressure within the laboratory is kept below that of the surrounding rooms and corridors, and stale air is extensively filtered before being exhausted. All samples and equipment, and the waste from sinks, are sterilised before they leave the laboratory.[4]

During the deliberations of the Working Party an extensive and detailed code of practice was constructed. The difference in effective containment between the four categories owes at least as much to the methods and practices involved as to the physical characteristics of the laboratory. In this light it is therefore significant that the chairman proposed in his initial paper that it might well be possible to adopt major aspects of existing codes of practice. Thus, although it was believed by some that the possible risks which might be associated with genetic manipulation experimentation could have involved some *novel* form of hazard, the immediate response of the Williams Working Party, as of Ashby and Asilomar, was to work on the basis that any risks would be comparable to orthodox risks, and to guard against them in like manner.[5] More specifically, the committee turned to the culturally available resources of the Godber categories of containment, and to other existing codes of practice.

These embryonic tendencies grew during the first few meetings. The general thrust was to press ahead with defining implementable degrees of containment (a concrete proposal on this was prepared by Brenner), whilst beginning to approach the question of the assessment of the possible risks of particular classes of experiment. This is dealt with in more detail below. Although, logically, it might seem that these two processes could not occur in isolation, participants' accounts stress the priority of fixing a set of containment categories based on existing practices, and *then* constructing a scale of possible hazards to match.

In tandem with these initial deliberations on the code of practice and the categories of containment, the Working Party also began to formulate its ideas on the organisational aspects of the Central Advisory Service (CAS). An early expectation was that whilst any experiment involving dangerous pathogens would have to be subject to at least the precautions required by Godber, other experiments (in categories C and perhaps B2) might be judged by the research councils, who would be able to refer cases to the CAS where necessary. The precise position and responsibilities of the CAS are unclear in accounts of this period given by participants, but in general terms it seems that decision making would be split between the CAS and the councils. This would probably have meant that the CAS would be responsible for setting broad standards, whilst the research councils would exercise their judgment within this framework. In effect, the CAS would offer advice to the councils, not direct to scientists. Observance of the code would be ensured by the appointment of laboratory-based or institution-based biological safety officers. Similar officers are required in most codes of practice. This arrangement for the CAS owes much to the model of DPAG established on the recommendation of the Godber Working Party. DPAG tenders its advice on the suitability of laboratories to undertake specified programmes of research on specified category A pathogens to government departments. The proposed CAS would, however, differ in that experiments would be considered individually, whereas DPAG, through government departments, effectively grants a licence for defined areas of experimentation.

It should be stressed that these ideas were only the first stage in a process of consideration, and that modifications were soon introduced. All the same, it should also be recognised that these early ideas carried the implication of scientists retaining a high degree of professional control over policy on genetic manipulation. As it happened, the idea of a judgmental role for the research councils appears to have quietly faded away.

Perhaps the first attempt to broaden the considerations of the Working Party came when Owen of the HSE pointed out the implications of the HASAWA, and the obligations of employers, including universities and research councils, for the health and safety of their employees, and of the public at large. He also made it clear that the TUC had intimated that they would be reluctant to back any code of practice produced by the Working Party unless they had been involved in its creation.[6] Owen's comments do not seem to have had any *immediate* effect, except that the Working Party agreed that its tentative conclusions should be circulated and followed by an extensive round of consultations with interested parties.[7]

At about the time of its second meeting in November 1975, the Working Party began the process of structuring its ideas about the organisation of the proposed system of advice. There was felt to be a need for some intermediate layer between the biological safety officers and the CAS. By analogy with existing MRC arrangements in other fields, a panel of some twenty referees responsible for screening proposals for genetic manipulation experiments was proposed. The panel might, for example, have the authority to agree on the categorisation of lower risk experiments and allow them to go ahead, whilst higher category experiments would go to the CAS. Whilst this could be justified on the grounds of organisational efficiency, and in terms of aiding communication, it would also have increased professional control over the assessment of experiments. At the same time it would also have confirmed the central body in an advisory role and have left the matter of ensuring observance of the code of practice in the hands of scientists – the biological safety officers and the panel members. But, contrary to the implication of this description, it should be recorded that the Working Party were, at this stage, considering the question of how to ensure compliance. Consequently, the chairman and the secretariat agreed to produce a paper outlining ideas on the organisation of a Central Advice *and Control* Service (CACS).

Towards the end of 1975 the Working Party discussed the paper on the CACS. On the assumption that it would be possible to categorise experiments in terms of the actual or potential hazard they presented, the paper anticipated that the CACS would have three functions: to advise on the categorisation appropriate for particular experiments; to advise on the local implementation of the code of practice; and to review scientific developments pertaining to genetic manipulation so that appropriate changes in the code of practice and categorisation could be made.[8] To fulfil these functions, and in order to be able to command the respect of scientists, it was felt that the central body would need to consist of eminent scientists with expertise covering genetic manipulation techniques, safety issues and industrial applications. It was proposed that the function of the former panel of referees should now be undertaken by local advisors who would also be members of the central body. They would be able to offer advice and be able to allow intermediate category experiments to go ahead. Local advisors would also be able to keep in touch with developments in their area. This would facilitate the gathering of information, and also allow them to act as a sort of 'inspectorate'. Local safety committees would be established, along with biological safety officers.

This three-tier structure would work as follows: the initial consideration of a proposed experiment would take place within the local safety committee, who would then assess it as a category A, B or C. (This short-lived tripartite classification of experiments into categories A, B and C was soon to be replaced by a more settled classification of A, B, C and D; with A being the most stringent and comparable to Godber/DPAG category A.) If it were clearly the lowest category, C, the experiment could go ahead. The local advisor would be consulted on category A and B experiments, or wherever there was doubt about the appropriate classification. The decision would be based on expert knowledge of categorisation, and on detailed local knowledge of conditions in the particular laboratory. This would include, for example, the adequacy of the equipment and the training of the people involved. The local advisor would be empowered to agree category B experiments. All category A experiments, those that involved novel techniques, or those which it was proposed to advise *against*, would be referred upwards. The CACS, in considering these high risk experiments, would have powers to accept or reject them, or to insist on specified precautions being taken.

The paper produced by the chairman and the secretariat also raised a series of questions, though without formally proposing any answers. First, 'Should the central body be in the business of advice or control?'. Though the title of the paper was 'Central Advice and Control', the text pointed out what must have been obvious, that the distinction between the two was becoming increasingly blurred. The paper went on to rehearse the argument of the Ashby Working Party for a system of voluntary registration and control, and the Williams committee's remit that it should consider the practical aspects of applying the Godber Working Party's recommendations. This led to the second and consequential question of whether the proposed functions of the CACS should be performed by an enlarged DPAG. Since the latter's control was presently based on an interim system of voluntary compliance pending a statutory basis, this raised the question of whether specific legislation should be recommended to cover the registration and/or control of genetic manipulation experiments.

A number of observations are in order here. First, despite the neutrality of the paper on the issue, it is difficult to envisage how the proposed structure of the CACS could have been grafted on to DPAG without fundamental changes in one or the other. Indeed, the paper cited an earlier Royal Society comment (Royal Society, 1975) that a small DPAG was incompatible with being able to give adequate consideration to the issues of genetic manipulation. Second, and in addition to these organisational factors, the Oxford conference had already produced arguments for distancing issues of genetic manipulation from those of dangerous pathogens. Thus the consideration seems to have been something of a ritual. Having fulfilled its charge to 'consider the practical aspects', the Godber/DPAG alternative was rejected by the Working Party. On the third question of whether there should be separate legislation to cover genetic manipulation, the committee felt at this stage that it did not need to consider the matter, although this view subsequently changed. It did, however, note that there would be great pressure on laboratories to follow the code of practice of the CACS, since failure could, in a case before the courts, lead to prosecution under the HASAWA. The general drift of this set of policy decisions implies a move away from the idea of formal control. In line with this, the name of the proposed central body reverted to the CAS.

An important and long-running issue which appeared in the chairman's paper was a reference to the kinds of criteria which would be used in the assessment of experiments. In its initial considerations of a proposal, the local safety committee would discuss both the scientific merits and the potential hazards of the experiment. Similarly the CAS would be able to reject a proposal if it felt that the potential hazards outweighed the potential benefits. This echoes Brenner's suggestion, in evidence to the Ashby Working Party, that scientific quality should be a factor in the assessment of proposals to perform genetic manipulation research. Related though slightly ambiguous statements about scientific merit found their way into the final report of the Working Party and eventually became incorporated in an advice note to scientists issued by GMAG. This caused great concern in certain scientific quarters since it clearly licensed an evaluation of the scientific worth of an experiment. In the later GMAG context, this meant that non-specialists, including scientists from other areas, trade union representatives and laboratory assistants, would be called upon to exercise a form of peer review. The scientific merit clause was also modelled on a statement in the report of the Godber Working Party, of which Sir Robert Williams was a member. Commenting on the factors to be taken into account in decisions about the use of highly dangerous pathogens, it advanced the principle that:

> The development of a technique, the educative value of
> carrying out a process, the desirability of keeping specimens
> for reference, general *scientific merit* or even the probability
> of a specific contribution to knowledge would not of

themselves justify automatic acquiescence with a request to use category A pathogens. The risk to the public presented by pathogens must be balanced against the possible benefits to the public to be derived from work on them. (Godber, 1975:11, emphasis added)

Translated into the context of genetic manipulation, the scientific merit clause could be taken to imply that an assessment of the potential scientific worth of a proposed experiment should be discussed. When it was first considered by the Williams Working Party the scientific merit principle was somewhat less radical than it later appeared, since the question of inclusion of non-scientists on safety committees had not yet arisen. Additionally the proposal may have seemed reasonable in the light of the then current fears about genetic manipulation. Surprisingly, those interviewed have no clear recollection of the reasons for the acceptance of this scientific merit clause.[9]

Early in 1976 the Williams committee produced the first draft of its report. The section on the CAS had apparently changed little from the proposals described above. However, a second draft, also early in 1976, contained a significant addition. It was now considered that because of the severity of the potential hazards the government should have appropriate powers to ensure compliance with the advice of the CAS. Further, it was suggested that existing legislation could not provide the necessary framework. The HASAWA, in particular, was now deemed deficient as its provisions only covered human beings, and not plants and animals, where the potential hazards were perhaps greater than for humans. New legislation would therefore be required to ensure that scientists, or laboratories, would have to be registered and licensed to perform genetic manipulation experiments.

Once again the available resources of the Godber report and DPAG are important here. In this second draft report the hazards of genetic manipulation were seen as potentially comparable with those of dangerous pathogens and as requiring a parallel system of control. DPAG then operated an interim voluntary system of control which would eventually be covered by specific legislation. This had been recommended by the Godber Working Party who felt that existing legislation was inadequate. In particular the HASAW Act did not provide for 'a convenient and quickly responsive system for the control of hazards which are unlikely, but could be capable of causing extensive harm' (Godber, 1975:19). Thus, the second draft report of the Williams Working Party suggested specific legislation to create power of 'registration and control' in line with Godber. The CAS would, however, be separate from DPAG since it was recognised that the addition of expertise to deal with genetic manipulation would have enlarged its membership well beyond the point of reasonable efficiency.

In subsequent discussions the Williams Working Party largely overturned this view on the necessity of legislation. The HSE felt that the HASAWA did indeed provide adequate powers of control, even if it only covered hazards to humans. Scientists felt that the advice of the CAS would be authoritative without the need for any statutory backing. In any case, legislative control would not be sufficiently flexible to cope with the situation of an ill-defined hazard where the rapid development of knowledge could soon render controls obsolete. Clearly, neither the interests of the HSE nor those of the scientists would be well served by the introduction of new legislation. It seems that the MRC was also against the idea of specific legislation. It felt that (a) the HASAWA already laid a clear obligation on employers, and (b) it would be difficult to justify legislation for a merely potential or even hypothetical hazard when other, actual, biohazards were not covered. If such proposed legislation were to have gone ahead the MRC, and hence the DES, would have been placed in an embarrassing position.

Other developments that emerged from the Working Party's discussion of the second draft of its report were that local safety committees would now be responsible for the assessment of proposals that fell into the two lowest categories (C and D). The treatment of the higher categories (A and B) would be as before, with local advisors able to deal with category B. The importance of the local safety committees was now given greater prominence than before. Finally, the Working Party agreed that in order to provide for a fast response the CAS would tender its advice direct to scientists and not, like DPAG, through a government department. The HSE would be notified of all proposals.

During the period in which the second draft was being considered and revised the Williams Working Party was beginning to receive the first responses in its consultation procedure. Indeed, the discussions on, and overturning of, aspects of the second draft were almost certainly influenced by the early results of the consultation. A consultative document had been issued which contained an outline of the committee's thinking at about the stage of the first draft report. In relation to the CAS it covered such points as the need for an eminent, specialist advisory service with a status comparable to that of DPAG. Scientists would be *expected*, but not required, to submit proposals, and to follow the advice on categorisation. The three-tier system was outlined, including the local safety committee's consideration of the scientific merits of experiments, and the proposal that the CAS would take into account the potential benefits of experiments, as well as the potential risks. No mention was made of the HSE or the HASAWA.

A good deal of written evidence was submitted during March and April 1976 by individuals and organisations. Oral evidence was also presented by trade unions, the Confederation of British Industry (CBI) and the Committee of Vice-Chancellors and Principals (CVCP). In general there was wide agreement on the need for some central body along the lines of the Ashby recommendation, though opinions varied on the details of its function and status. The great majority of the evidence from individual scientists,[10] and from scientific bodies such as the Institute of Biology and the Royal Society, concentrated on such areas as the code of practice and the four containment categories, especially on their stringency in relation to the NIH guidelines. There was no adverse comment on the recommendation that local safety committees should discuss the scientific merits of proposed experiments, and only one or two comments to the effect that the recommendation concerning the CAS implied an illegitimate extension of the peer review system.

The surprising lack of adverse comment on the function of local safety committees, like that on the Working Party itself, was perhaps due to the fact that no details of the membership of the local committees were then available. However, similar considerations applied to the CAS, and objections to the illegitimacy of the implied extension of peer review in this area did not lead to a corresponding rejection of the scientific merit clause for local safety committees. Once again, it may have been that the potential hazards were thought to be so severe that some such assessment was in order.

Scientists took readily to the Working Party's proposals for a voluntary system of control and there was no call from scientists for any statutory basis for the activities of the CAS. Similarly, the conceptual and organisational separation of genetic manipulation from dangerous pathogens was well received. The main area of scientific concern was that the advisory body should be established as soon as possible, and its operation should not delay the performance of experiments.

The response of the trade unions was perhaps equally predictable. In oral evidence the TUC representatives, including ASTMS and NALGO members, the Association of University Teachers

(AUT), and the Institute of Professional Civil Servants (IPCS), all emphasised the strong need for some form of legally enforceable controls on genetic manipulation. The TUC representative put forward a policy based on five points:[11]

(1) enforceable regulations under the HASAWA,
(2) the CAS should be an agency of the HSE (to provide the latter with scientific expertise),
(3) trade union representation on the CAS,
(4) the CAS should maintain a register of all genetic manipulation work, and
(5) trade union laboratory safety representatives should agree all proposals.

The AUT and IPCS broadly supported a role for the HSE in the control of genetic manipulation, but in other respects their policies do not seem to have been worked out in terms of such detailed goals as those of the TUC. The involvement of the HSE in the area was also acceptable to both the CVCP and the CBI, though in neither case could it be termed a policy objective. The CBI also felt that there would be no great difficulty for industry in providing the kind of information that the CAS was likely to require.[12]

The most detailed and extensive comments submitted to the Working Party came from the MRC. The opinions presented in the MRC evidence were canvassed at a small meeting in March 1976 of senior scientists which included several of those sitting on the Working Party. To summarise from the many comments made in the MRC evidence, it was argued that confusion would be caused by labelling the containment categories in a way that suggested a relationship to those of the Godber report. Following this the labels were changed from A, B, C, D, to IV, III, II, I, where I was the lowest level of containment. The MRC evidence also seems to have been based on a very clear idea of what the status and powers of the CAS should be. Obviously reacting to the later, second, draft report rather than the Working Party's consultative document, the MRC evidence argued strongly that the CAS should not have statutory powers to prevent experiments being performed. Rather, it should be an advisory service on categorisation and codes of practice. Judgment of the scientific benefits of experiments by the CAS was also rejected, as was the intermediate, local advisor, level of the three-tier system of advice.

In response to the consultative document itself, the MRC felt that some rather woolly language should be tightened up, to make it clear that all scientists would submit details of experiments to the

CAS.[13] This would then constitute a registration system. Copies of the register would be sent to the HSE, thus rendering specific legislation requiring registration unnecessary. In view of the onus placed on employers by the HASAWA it was argued that legislation to *control* genetic manipulation was also superfluous.

Following the period of consultation, a third draft report was produced in April 1976 in which the CAS finally became known as the Genetic Manipulation Advisory Group. In brief outline, a system of *advice* would operate through local safety committees who would be able to agree category I and category II (previously category D and C) experiments. Safety committees now had to be 'properly constituted and representative', though no detailed specification of this phrase was given. They would still discuss the scientific merits of experiments. The intermediate layer of local advisors was abandoned, all category III and category IV proposals, and notifications of agreement to category I and category II proposals, would now go direct to GMAG, who would *not* engage in any review of the potential benefits of experiments. Rather, they would offer advice on categorisation, perhaps direct to scientists, or perhaps via a government department. The Working Party had again changed its mind on this point and now left the final decision to the government.

On the question of control, the third draft came to the conclusion that, given the background of the HASAWA and that GMAG would pass on copies of its register of proposals to the HSE, laboratories would, in fact, conform to this 'voluntary' system of control. Therefore, specific legislation was not required. Thus at this stage the Williams Working Party had effectively adopted the MRC position, which was broadly compatible with the general tenor of comments from individual scientists, that voluntary control was acceptable. Though the HSE would be involved it was by no means the close relationship outlined in the TUC objectives, for no statutory control under the HASAW Act was recommended. Neither was there any recommendation on trade union membership of GMAG or the local safety committees. The Working Party did, however, point out that the government could, in the light of experience of the operation of the voluntary system, later consider the advisability of additional, statutory measures.

Most of the detail of the third draft remained unchanged and appeared in the published report of the Working Party. Some aspects were, however, revised as the report went through a fourth draft, where it was felt that GMAG should tender its advice direct to scientists, and that the group should be run by a government department. Later, this was once more changed. In the published report GMAG was given more independence, and the connection with a government department loosened to the extent that the latter would be involved only in providing a secretariat.

On the subject of control the fourth draft was compatible with the third, but at the very last meeting of the Working Party, in May 1976, it was agreed that the published report should positively recommend the introduction of regulations under the HASAWA to require laboratories to submit proposals for all experiments to GMAG and the HSE. Thus, in the end, a statutory system of *registration*, but not of *control*, was recommended. In the report the Working Party also rehearsed very briefly the arguments in favour of control under HASAWA, the counter-arguments concerning the lack of applicability to plants and animals, and the possibility of consolidating legislation as envisaged by the Godber report for dangerous pathogens. The final choice was left to the government to make in the light of experience with the voluntary system.

The final, significant, change to note was made *after* the last meeting of the Working Party. The various drafts had referred to the 'broadly based' membership of GMAG in terms of scientific eminence, and the expertise required, but the report, published in August 1976, has it that

> the membership of the GMAG should include not only
> scientists with knowledge both of the techniques in question
> and of relevant safety precautions and containment measures
> but also individuals able to take account of the interests of
> employees and the general public. (Williams, 1976:13)

As at other points in the development of the British policy on genetic manipulation it is difficult, if not impossible, to be certain of any specific influence which might have led to this particular recommendation being included. It seems extremely unlikely that such an important policy initiative came from within the Williams Working Party.

When the TUC delegation had presented oral evidence, including a demand for representation on GMAG and safety committees, they were left with the impression that the Working Party considered such matters to be outside its remit.[14] Members of the Working Party interviewed seemed uncertain about the recommendation's origin, or were unwilling to go into details. However, given that ASTMS felt that they had been assured by Fred Mulley, Secretary of State for Education and Science, that, in line with their objectives, they would have some part to play in the implementation as well as the development of genetic manipulation policy, it may well have been a political decision. A number of participants felt that this was the case, though they all attributed the initiative to a later Secretary of State for Education and Science, Shirley Williams.[15] In fact she did not replace Fred Mulley until September 1976, and after the publication of the Williams report. Whilst Shirley Williams was very closely involved in decisions about the eventual numbers of the representatives of the interest groups, and the individuals chosen to fill these roles, the decision on the principle of their inclusion, which makes GMAG such an interesting innovation, had taken place earlier.

This section has described in some detail how policy was evolved, but occasionally has been unable to point to specific 'causal' influences. In part it is a consequence of fine-grained analysis that questions which have not previously been raised are posed. But there are problems in attempting to analyse policy making which takes place behind closed doors. Many of the participants interviewed also had some difficulty in identifying influences, or in one or two cases were unwilling to do so.

Of course, when participants were attempting to provide a rationale for policy decisions there was a tendency to present those decisions with which they agreed as having emerged naturally from extensive discussions and from the logic of the situation. Those with which the speaker disagreed, or which were viewed with suspicion, were often, but not always, presented in terms of a reconstruction based on the known or perceived objectives of certain individuals or groups. The general problems that this raises for the social scientist have been well explored.[16] In this account an attempt has been made to indicate the more important uncertainties. In practice, there was a high degree of emphasis on the naturalness of the decisions taken, reflecting the homogeneity noted in the introduction to this chapter that the members of the Working Party were drawn from the senior scientific establishment who interacted both with each other and with the MRC.

The principles for the assessment of risk

When the Williams Working Party began sitting in October 1975 the process of the construction of the American NIH guidelines was well under way. The Recombinant DNA Advisory Committee had already produced the Hogness and Wood's Hole versions of the guidelines. These were followed in November by the Kutter version and in December by an important conference in La Jolla, California, where the various versions were exhaustively compared, and from which a set of more firm proposals emerged. Further consultations led to the publication of the 'final' version of the voluntary NIH guidelines in June 1976, at about the time when the Williams report was being prepared for printing. Further revisions to the NIH guidelines began almost immediately after the publication of the 'final' version. The need for revision had been recognised prior to June 1976.

In abstract, the technical problem for both American and British scientists, and others concerned with the detailed development of policy, was the same: to produce a scale of potential hazards which could then be matched to a scale of appropriate precautions.[17] The major difficulty was that the hazards were putative or hypothetical since there was little or no direct evidence available that could be brought to bear on the issue. Allied to this, it can now be seen with the benefit of hindsight that there was, initially at least, a problem of lack of understanding between scientists in different areas of biological research. Molecular biologists seem not to have fully grasped the meanings and implications of research on, for example, the ecology of *E. coli*, and of known pathogenic organisms. Similarly, those concerned with these areas of research were uncertain of the detailed implications of the new genetic manipulation techniques.[18]

In general, then, there was an area of scientific uncertainty in the face of the powerful scientific and social pressures to ensure that the research was performed safely. It was explicitly recognised by the organising committee of the Asilomar conference in their report to the National Academy of Sciences that there could be no properly 'scientific'[19] assessment of risks based on data, but that the consensus of experts would have to be taken as the best available source of information. This assumption had also been implicitly made by the Ashby Working Party.

Johnston (1980) has characterised the research area of risk assessment as the transformation of 'uncertainty' into 'risk'. Fundamental uncertainties about both the knowledge-base and the role of human values are routinely hidden or avoided by a reliance on the often inappropriate conceptual and technical apparatus of science. Citing endemic conflict and lack of consensus over whether risk can be operationalised or measured, Johnston argues that risk assessment is paradigmatically immature, and also relies on inappropriate probabilistic models imported from physics (see also Wynne, 1975).

In the particular case of genetic manipulation, no professional group of risk-assessors had any experience or deep knowledge of the new techniques. Moreover, there were no known risks to be measured, even by 'immature' or 'inappropriate' standards. Thus the process of turning uncertainty into risk in the case of genetic manipulation was, paradoxically perhaps, less open to charges of scientific sleight of hand. The application of the patina of science was even more problematic than Johnston implies is usual in risk assessment. This is not to suggest that molecular biologists acted out of pure self-interest or that the regulation of genetic manipulation was arbitrary or unconstrained by scientific argument, but rather to underline the very limited extent to which scientists expected a proper scientific solution to the problems of recombinant DNA research.

In the American approach to the construction of guidelines, and more particularly the scaling of hazards, a set of four physical containment levels (P1 to P4) and three biological containment levels (EK1 to EK3) was eventually settled on. In broad outline, and somewhat idealised, the procedure was then to construct an extensive table of all possible experiments (or classes of experiments) and to allocate them to appropriate physical and biological containment levels. Tooze (1978) has characterised this as an 'encyclopaedic approach'. Allocation was based on a consideration of the classes of experiment, and was carried out in such a manner that the detailed arguments about each experiment or class were potentially open to scientific and public view, and influence. This can be thought of as similar to a codified system of law.

The details of the NIH system and the process of developing the classes of experiment were sufficiently well advanced for them to be known to the Williams Working Party in October 1976. Brenner, a member of the Working Party, had been involved in some of the developments in the United States, and attended the important La Jolla conference. It would have been perfectly open to the Williams committee to have adopted any of the hazard or containment categories, or the assessment scheme eventually to be produced by the NIH. Indeed, several scientists suggested in evidence to the Working Party in the spring of 1976 that they should so do.

Why the Working Party chose to follow an independent path is not known for certain. National independence and pride were significant, but not determining factors. One aspect may have been the variations in the different versions of the NIH guidelines. Probably a more important factor was a fear that social and political pressures in America, such as those which led to the establishment of the Citizens' Review Board in Boston and the moves in the US Senate for legislation, would lead to strict legal controls on genetic manipulation. Should this be the case there would be obvious advantages in having an independent system. In addition, Brenner and perhaps another scientist, Walker, who had been able to follow the American developments closely, were firmly of the opinion that the US system of codified law was far too inflexible. Should it become enshrined in British law, or even only in accepted codes of practice, it would prove more difficult to change than in America. A system responsive to changes in scientific information about the potential hazards was an important consideration for British scientists.

By contrast with the codified law of the American system, the approach finally recommended by the Williams Working Party was based on a case law philosophy. In its purest form, in 1976, the case law approach proposed that the CAS should exercise its expert judgment on the basis of certain rather general principles and paradigmatic classifications. In the published report of the Working Party a case law approach was still an important recommendation in formal terms, but its effect was somewhat diluted by the inclusion of a more extended table of paradigmatic classifications.[20]

The Williams committee gave precedence to physical containment over biological containment, on the grounds that the former was well understood, whilst the latter was a novel concept whose implications and effectiveness were not yet clear.[21] This contrasted with the optimism of the Ashby Working Party and that enshrined in the NIH guidelines. In the table of suggested paradigmatic classifications in the Williams report, a limited role was given to biological containment, since the use of a disabled host/vector system carried a level of containment which was one category lower than if a non-disabled system were used. The Working Party did not, however, specify what was to count as disabled, preferring to leave this task to GMAG.

The Williams committee started its work by considering a paper from the chairman proposing four categories of containment, the higher of which were modelled on the recommendations of the Godber Working Party. At this point it was proposed that the CAS should be given only very limited guidance on the categorisation of risk, and be expected to undertake the development of the more detailed case law approach itself. Almost immediately, however, the need for somewhat more detailed guidance for the CAS was accepted, although the principle of the case law approach was retained. The Working Party proceeded by first defining implementable degrees of containment, and then attempting a classification of potential hazards that would be sufficient to enable the CAS, later GMAG, to begin its work.

The task of producing an initial classification fell in particular on three members of the Working Party: Brenner, who has already been mentioned in several contexts, Walker, then Director of the MRC Mammalian Genome Unit at the University of Edinburgh who was one of the 'scientists working in Scotland' noted in chapter 2 in connection with evidence to the Ashby committee, and Tyrrell, a scientist at the MRC Clinical Research Centre at Harrow and the MRC representative on the Godber Working Party, and who was at the MRC meeting which considered the interim Godber report just prior to the publication of the 'Berg letter'.

The first proposal for an assessment of possible hazards was due to Walker and Tyrrell. It appears to have been based on their feeling that it was possible to produce a rank order of hazards that would command a wide consensus. A set of headings covering the aspects of an experiment which it was necessary to take into account in assessing any hazard was also given. The set was essentially similar to those produced in the United States. The headings were: the source of DNA used in an experiment; the nature of the particular fraction used; the nature of the bacterial host; and the type of manipulative procedures used. Within each of these headings various factors were ordered in terms of the severity of possible risk and also scaled using a simple points system. The sum of points or the score of any experiment was then a rough measure of its hazard; this could in turn be matched to a scale of containment levels, for example 0 to 4 points, category C; 5 to 9 points, category D; etc. Obviously, this goes beyond a rank ordering, since some minimal assumptions are made about the relative risk values of the various factors in order to allocate points to them.

It seems that there were too many anomalies and exceptions to this scaling system, and the points approach was soon abandoned. The principle of rank ordering was however preserved in the published report of the Working Party, though there were some changes in content and in presentation. The main change was that the headings concerned with the nature of the plasmid or virus vector and the nature of the host organism were amalgamated under one heading – the host vector system. Otherwise the scheme was essentially that found in the published report.

In the Walker and Tyrrell proposal a set of illustrative examples of categorisation had been given. Following the abandonment of the points system the amount of detail in the examples was increased, and a certain amount of consolidation took place, though this does not seem to have resulted in any significant change in the assessment of individual risks. The set of examples was eventually published as the table of suggested categorisation in the Williams report. Along the way various adjustments had been made to the table as the committee had opted for three levels of containment, and then returned to four. Marginal changes had been introduced in response to the criticism voiced during the consultation process, and the limited extent of biological containment had been clarified. In general, then, there was some tinkering at the margins of categories which probably resulted in the re-allocation of a few experiments, and details, merely implicit in earlier formulations, were more explicitly dealt with later on. This had the effect of improving the expression and changing the appearance of the Working Party's recommendations, but it does not seem to have involved much substantive change. Indeed, when considered together with the detailed arrangements specified for the operation of the four containment levels (the code of practice) there seems, overall, to have been remarkably little change in the perceptions of the rank order of the severity of possible hazards during the life of the Working Party. This raises two sets of questions about the principles of the assessment of risk used by the Williams Working Party: the level of the perceived hazards vis-à-vis the American guidelines; and the basis on which the assessment of risk was made.

On the first point, comments received by the Working Party during the consultation process made it clear that a good number of scientists thought the British proposals were more strict than the NIH equivalent. This was especially true of category III, which was seen as considerably more stringent than the equivalent NIH category P3. The Working Party appears to have taken the view that, overall, the differences in severity between the British and American levels of containment were not particularly significant. Certainly no experiments were positively prohibited under the Williams system, unlike the NIH guidelines. Also the Working Party, and in particular Brenner and Walker, felt that the important point about the Williams remit for GMAG was the proposed flexibility enshrined in the injunction to establish its own case law. On this view any difficulties or anomalies could easily be dealt with as they arose.

On the question of the principles of risk assessment it is worth emphasising the fact that both British and American scientists felt able to rank order the possible hazards. In the absence of any direct experimental evidence this assessment was based on the scientists' background tacit knowledge which, as Polanyi (1958) has argued, is difficult, if not in principle impossible, to explicate fully (see also Ravetz, 1973). This is not to say that reasoned argument is impossible here, but rather that arguments and explanations offered may have the character of *ex post* justifications. To be more specific, it seems that the early points system, for example, should be viewed as a means of displaying and presenting scientists' implicit categorisations or tacit knowledge about the hazards. In the final analysis the points system was dependent on the tacit knowledge of the scientists and not the other way round, even though the associated numbers make it appear that the categorisations are the result of, or derived from, independent, quantified evidence. This kind of reversal of the apparent logic is, of course, fairly common in the presentation of science as, for example, Medawar has argued in his 'Is the scientific paper a fraud?' (Medawar, 1963). Such considerations largely explain the lack of real change in the assessment of hazards; the perceptions were firmly anchored in the solid and extensive base of tacit knowledge.

Related to this general point there is a particular and consequential aspect of the Williams approach to the assessment of risks that will now be examined in more detail. This is the factor of 'phylogenetic relatedness' which is implicit in the table of paradigmatic examples in the published report, and which is also more explicitly stated in the text under the heading 'Source of nucleic acid', part of which reads:

> Nucleic acid from higher organisms is generally considered to offer a hazard related to the closeness of the evolutionary relationship between the organism constituting the source of the nucleic acid and the organism at risk. On this basis we should regard nucleic acid from plants and invertebrates as carrying a low hazard to man (except perhaps for invertebrates that may harbour microbes pathogenic for man); nucleic acid from amphibia, reptiles and birds would be assigned an intermediate hazard classification, and nucleic acid from mammals, including man, would be the most hazardous. (Williams, 1976:6)

The reasoning which lay behind this proposal, and a similar principle in the NIH guidelines, is now difficult for scientists to reconstruct. It was often described in interviews as a 'mistake', or as having 'no foundation'. Tooze has suggested that:

> The rationale for this [principle of phylogenetic relatedness]

was that the more closely related the two species are [the donor and recipient of the DNA] the greater the homology between their DNA sequences, and therefore, the greater the chance of a recombination event between the two DNAs. Furthermore it was argued that since the host range of many viruses appears to have a phylogenetic basis, the possibility of inadvertently activating a latent virus of the donor species that could infect the species at risk was greater the closer the phylogenetic relationship of the two species. (Tooze, 1978:281)

It is difficult to judge just how reasonable these arguments were at the time. Obviously they must have had some *prima facie* plausibility, and they should not be dismissed just because they are now, in the light of more recent views, seen as simplistic. But even allowing for the benefit of hindsight it does seem that the principle of phylogenetic relatedness was remarkably undetermined by the then available evidence. Why then was it used?

One short answer in the case of the Williams committee was that the principle was implicit in NIH guidelines, and it was thought that the British version should not be too different. However, this explanation does not help greatly, since the Williams committee could perhaps have adopted some other principle which would have had a similar effect in terms of categorisation. In any case, one would have to ask why the NIH guidelines had adopted the principle in the first place. This in turn devolves on to a question about the structure of knowledge in biology, and how it affected both the American and the British approaches to the construction of guidelines.

In contrast with popular conceptions of physics as having a core of formal quantitative theory, biological knowledge is characterised by a series of less formal models.[22] As basic perceptual and conceptual tools, these models structured biologists' thinking about the possible outcomes of genetic manipulation processes. In particular, consideration of the effects of mixing genes from different organisms would automatically involve the basic model of the phylogenetic scale and the associated evolutionary principles. The biological characteristics of organisms are routinely related to each other in this fashion, which is roughly comparable to the manner in which chemists relate the properties of elements through the periodic table. That is, the model structures and organises thought and presentation; any particular characteristic can be related back to the model. However, the phylogenetic scale model is not in any simple sense predictive, and there are always very many special factors to be taken into account in any explanation. Indeed, biological knowledge in general, and molecular biological knowledge in particular, requires its practitioners to be familiar with a vast range of detailed background knowledge about particular processes, in particular organisms, in particular situations, and so on. The context-dependency of this knowledge, or to put it another way, the necessity for taking into account potential anomalies when interpreting general principles, is a routine matter for biologists. Fellow biologists recognise the implicit use of *ceteris paribus* or *mutatis mutandis* clauses when discussing areas of accepted knowledge. Thus models and the principles implicit in them, and the background knowledge of particular cases, can be maintained as a coherent whole.

In the case of genetic manipulation, however, there was no accepted knowledge on the specifics of potential risks, and hence the coherence of facts, models and principles was potentially in danger.[23] This did not have any great consequence, at least initially, for the American approach to construction of guidelines. Because of the 'encyclopaedic' approach, the Americans were able to obviate the anomalies in the application of the phylogenetic principle by incorporating background knowledge into decisions about particular cases. This had the corollary of laying their proposal open to charges of being *ad hoc*. That is, the phylogenetic principle was recognisably present in the American guidelines as a feature of, but not as a rule of, decision making.

In contrast, the Williams Working Party seems, after the initial points system, to have intended a case law, rather than an encyclopaedic or codified law, approach. In an ideal-typical sense of a case law approach, cases are logically prior to principles. Local and finite principles are expected to emerge from the consideration of particular cases. However, the Williams committee felt that some initial guidance for GMAG was in order, and thus principles for decisions about the formulation of case law were laid down. The apparent intention of the Working Party was to provide a simple and limited set of principles with which GMAG could begin its work, and it was not expected that the principles would be taken too seriously, or followed for too long. GMAG did not perform in the way that at least some members of the Williams committee had expected it would. The explicit formulation of the principle of phylogenetic relatedness enabled, and perhaps licensed, GMAG treating it as a rule of decision making. In summary, then, the American and British approaches were based on a common structure of tacit knowledge, but reflected divergent modes of systematising that knowledge into a decision making procedure.

GMAG: from recommendations to practice

The Williams report (1976:iii) had noted that there was a 'pressing need for the implementation of a system of advice and control' which could be best met by a 'flexible approach' rather than through 'rigid guidelines' (Williams, 1976:4). Advice and control should be voluntary, but the Working Party recommended that the notification of experiments to GMAG should eventually become compulsory. This would be achieved by means of regulations made under the Health and Safety at Work Act. Further recommendations were that GMAG should be separate from DPAG and, in line with this, that it should deal directly with scientists rather than following the DPAG model of tendering its advice to a government department. Though GMAG should in some sense come under the aegis of an unspecified government department, it was clearly intended that it should have greater independence than DPAG.

Of course, these proposals were the outcome of a series of negotiations. It is important to recognise that whilst the Williams recommendations were significant indicators of agreement and/ or compromise, the process of negotiation extended beyond the Working Party, both in terms of the area of negotiation, and in time. Senior scientists, civil servants, members of the Health and Safety Executive, trade unionists, industrialists and politicians were all involved in a series of informal contacts and exchanges. During its life, the Working Party was the major focus of this activity, but it was not the only one, although information about influential decisions reached as a result of these external negotiations is scarce. It has already been noted, for example, that the proposed inclusion on GMAG of people able to take account of the interests of employees and of the general public came very late in the life of the Williams Working Party. This was not prefigured by the lengthy discussions which attended the other recommendations.[24]

In interviews those in a position to know the details of any such events were somewhat reticent, but did agree that as a general rule, verbal or more formal reports of the committee on certain significant topics would have been made to higher levels within government bodies, and any feedback duly noted.[25] The business of informal contacts did not, of course, cease with the writing of, or the publication of, the Williams report.[26] Indeed, negotiations about the role, functioning and location of GMAG continued until it was reconstituted in 1984. The remainder of this chapter will look at the important factors which shaped the initial structure of GMAG.

One significant aspect of the Williams report is those areas on which the committee did not make recommendations, or which it preferred to leave open. Perhaps the most important was the lack of specification of the relationships between GMAG and other government and official bodies, particularly the Health and Safety Executive. Full-time trade union officials and employers' organisations were both in favour of the genetic manipulation issue being dealt with by the HSE.[27] For both parties the advantage of such an arrangement was that it would give them a fairly well-defined role within an organisational structure with which they were familiar. Scientists, on the other hand, had forcefully argued the case for flexibility. They saw the need for the continuous input of advanced scientific knowledge so that the advice and control could keep pace with the expected rapid change in the field of genetic manipulation. They had strong reservations about the technical competence of the HSE. The more closely GMAG was associated with the HSE, the more the recombinant DNA issue would become assimilated into the area of industrial health and safety, and the less it would be a matter for professional scientific control.

The scientists' arguments on the need for flexibility are reflected in the Williams Working Party's recommendation that GMAG should not operate a statutory system like that of DPAG.[28] Here, the success of the scientists' view was probably aided by the HSE itself, which was in favour of an agreed code of practice rather than a formal statutory code. One aspect of the difference between the two forms of control is that failure to observe a statutory code would be sanctionable in itself, whereas failure to observe an agreed code of practice would be *prima facie* evidence of failure to properly protect employees' health in the case of an infection.

The issue for the decision making process, then, was one of striking a balance between the views of the trade unions, the employers, the HSE, and the scientists' favoured option of closer ties to, or control by, the DES, via the MRC.[29] The Williams report had narrowed the range of options, but performed only a limited adjudication between the various views. Although there is mention of the HSE, the relationship between it and GMAG is left unspecified. The Working Party had moved away from the Ashby philosophy of voluntary self-control by individual scientists, but did not go as far as putting the issue directly within the ambit of the HSE, as ASTMS, for example, would have preferred. The manner in which the relevant paragraphs of the Williams report (1976:16–17) are written seems to be rather carefully designed. Although the HSE would eventually produce a form of statutory notification, GMAG would not be too strictly bound to the HSE.

The HSE, it seems, would have preferred to have had a greater degree of responsibility for, or control over, GMAG. Indeed, the inclusion of a reference to a role for the HSE seems to have come rather late in the life of the Working Party, and, like the reference to representatives of the public interest, may well have been due to some external influence. The Working Party seems to have decided to avoid direct involvement in this and related issues, in effect passing the decision to the Secretary of State for Education and Science.

In outline, the discussions and negotiations over whether there should be voluntary or statutory control of genetic manipulation had begun before the Ashby Working Party, had continued through its deliberations, had passed to the Williams Working Party, and, especially towards the end of that Working Party's time, had involved active discussions in Whitehall. In this latter period, the scientists' option was expressed mainly through the MRC and thence to the DES, with the Ministry of Agriculture, Fisheries and Food, responsible for the Agricultural Research Council, playing a supporting role. The trade union position, and in context this means essentially the ASTMS position, had been pressed on, and was partly expressed through, the HSE.

No doubt there were subtleties in the two broad positions since the process was conducted largely behind closed doors, unlike the more open discussions in the United States. However, it remains true that the major competing definitions of the situation were those of professional self-control by scientists, and treating genetic manipulation as an industrial health and safety issue.[30]

One visible event in these negotiations about the relationships between GMAG and other official bodies, and hence GMAG's role and function, was the publication of a Consultative Document by the Health and Safety Commission. This contained a draft definition of genetic manipulation, and proposals for 'Compulsory notification of proposed genetic experiments in the manipulation of micro-organisms'. It was published at the same time as the Williams report, which had apparently been delayed in order to give the HSC time to formulate its approach (Lewin, 1976a). The wording and implications of the definition of genetic manipulation caused a storm of protest from scientists. It read:

No person shall carry on any activity intended to alter, or

likely to alter, the genetic constitution of any micro-organism unless he has given to the Health and Safety Executive notice, in a form approved by the Executive for the purposes of these Regulations, of his intention to carry on that activity. (HSC, 1976: Appendix B, para 2)

The first point to note about the passage is that it is concerned with compulsory notification, and does not imply that there will necessarily be any form of statutory control. That is, it was seen by the HSE as a means of ensuring that they, and hence GMAG, would be informed of such activity.[31] One problem with the definition is that it is very broadly drawn. As was forcefully pointed out to the HSE, the definition covered not only scientists' genetic manipulation experiments, but many long practised and unquestionably safe techniques in classical genetics, the medical use of X-rays, and even the treatment of garden rose bushes with pesticides and fungicides, both of which might alter the genetic constitution of micro-organisms. Perhaps worse, the definition, being restricted to micro-organisms, did not include those types of experiment involving the insertion of foreign DNA into higher organisms, namely plants and animals including humans.

In 'An open letter to the Health and Safety Executive', Ashburner (1976) articulated such points, and stated:

You remove the Advisory Group or any similar body from any central role in the dialogue that must exist between those who do the experiments and those who administer the laws under which they are done. With respect, sir, you do not have the standing in the scientific community required for this job. (Ashburner, 1976:2)

Ashburner concluded that:

It would be far more effective to draft the regulations to include only those techniques of *known* danger or which scientists judge to be of potential danger, and actively to involve both the scientific community and others in both the assessment of these dangers and the administration of the law itself. (Ashburner, 1976:3, emphasis in the original)

Criticisms of the Consultative Document and the definition of genetic manipulation were also made by a number of scientists and others.[32] Although the HSE could defend the logic of its position in terms of an intention to cover all possible forms of genetic manipulation and then to exempt non-hazardous areas (Locke, 1976), it was widely recognised that a serious mistake had been made.[33] The dangers of failing to gain the confidence and co-operation of scientists was clearly highlighted; the stock of the HSE among scientists, which was not, in any case, very high, fell.

The furore over the Consultative Document weakened the position of the HSE in subsequent negotiations. It seems safe to conclude that as a result GMAG ended up with a greater degree of independence from the HSE, and was closer to the DES than would otherwise have been the case. In line with this, any possible attempt by the HSE to propose a more inclusive definition of genetic manipulation than that in the Williams report was radically undermined.[34]

In the event GMAG was formally constituted as an independent body, a quango, with the responsibility of advising scientists and the HSE on a number of matters relating to genetic manipulation.[35] The DES was responsible for providing the secretariat through the MRC, and for the appointment of the members of GMAG. Thus, although GMAG had a measure of independence from both the HSE and DES, its formal organisational location was within an arena to which scientists had ready access. As both Yoxen (1979a) and Wright (1978) have noted, this location represented a victory for scientists seeking to maintain a channel of influence and control over the regulation of genetic manipulation.

Some participants expressed the view that the negotiations within Whitehall[36] about the role and function of GMAG, and hence its relations with other bodies, became quite strained. Several suggested that the matter had eventually to be decided at Cabinet level. Certainly, there was at least a limited amount of strain in the Whitehall negotiations, although the issue may have gone to the Cabinet partly because of the change in Ministerial responsibility for the DES which took place in September 1976 when Shirley Williams succeeded Fred Mulley. Also, it should not be forgotten that GMAG was a novel organisational response to the unprecedented character of the fears about the new genetic manipulation techniques.

A further input to the decision making process was a demand from ASTMS, made in August 1976, to the Secretary of State for Employment who was responsible for the HSC/HSE, that all laboratories should be brought under the HSE genetic manipulation regulations.[37] ASTMS were also in favour of GMAG being a delegated body under section 13 of the HASAW Act. This would mean that the statutory powers of the HSE would be delegated to GMAG, thus giving it some powers of enforcement.

It would also mean that, as a consequence, there would be a 50 per cent trade union representation on GMAG.

ASTMS had for some time been concerned about safety standards in laboratories, especially following the outbreak of smallpox in London in 1973. They had met HSE officials in June 1975 to discuss the Ashby and Godber reports, and had formulated a policy on genetic manipulation in the early part of 1976.[38] A significant aspect of their policy, as on safety matters in general, was an emphasis on the importance of involving employees and their unions in policy formation and watchdog roles. Their demands, then, represent a continuation of this general policy.

A related aspect of the negotiations concerned the representation of the various groups on the proposed GMAG. The main difficulty in this area was the number of trade unionists, or 'representatives of the interests of employees'. Initially it had been intended that there would be three, but comments from the TUC led to this being increased to four. Additional pressure was brought to raise this number to five, but the chairman-designate, Sir Gordon Wolstenholme, let it be known that this would be unacceptable as it would upset the balance of the committee.[39]

From a trade union point of view the upward pressure on the number of representatives on GMAG was a result of differences of opinion, particularly between ASTMS and the AUT. The latter felt that they should have a representative on GMAG, whilst the ASTMS and TUC view was that a significant number of AUT members would in any case be on GMAG as 'scientific and medical experts'.[40] Discussions on this issue, and parallel negotiations between other trade unions and the TUC, took some time, and led to the first meeting of GMAG, planned for November 1976, being delayed until December. Even then, the 'representatives of the interests of employees' had not been finally decided and the meeting went ahead without them.[41]

There were also difficulties over the status and selection of the scientists for GMAG. Whilst all agreed that the scientists should be eminent, it was more important that they should embody the broad range of expertise necessary for the group to function effectively. The official position of the DES was that their role and status on GMAG should primarily be that of the expert rather than the representative of a particular interest. In fact it seems probable that this 'official position' owed considerably more to the views of the then Secretary of State, Shirley Williams, than to her officials. Several scientific bodies such as the Institute of Biology made representations to the DES to be allowed to nominate members for GMAG, but these were refused.[42] Official consultations on the proposed scientific membership of GMAG involved the research councils, but not scientific societies (Select Committee, 1979:156 para 644). Apparently here, as elsewhere, Shirley Williams utilised her own personal contacts in the scientific community as a basis for a wider consultation.[43] The overall outcome of these sets of negotiations was that scientists were chosen on the basis of their potential contribution to GMAG, rather than as representatives of particular scientific societies, or solely on the basis of eminence.[44]

The third significant area of negotiation during the autumn of 1976 was the question of the terms of reference of GMAG. The new Secretary of State for Education and Science, Shirley Williams, took a close personal interest, and was very probably influential in, for example, the number and perhaps the choice of 'representatives of the interests of the public', as well as in the rather wider remit given to GMAG.[45]

During this period of negotiations over the shape and structure of GMAG, scientists' actions were mainly directed at preserving the spirit and important details of the Williams report against pressures for changes from the trade unions. Scientists were largely content with the Williams approach since it embodied a number of the policies adopted at the Oxford conference. These were: a rejection of the Godber committee's conjoining of the genetic manipulation and dangerous pathogens' hazards; the avoidance of special legislation to control genetic manipulation; the adoption of a flexible approach to categorisation of a limited range of experimental techniques; and the organisational location of GMAG within the ambit of the DES. As the negotiations had progressed through the Williams committee, the scientists had had to accept some modifications to the precise implications of these policies. These derived, in part, from the HSE's legitimate interest in the area, and from concern that groups other than scientists should have at least a hand in detailed decisions about possible hazards and appropriate precautions. One important objective of the Oxford conference had been the speedy introduction of a system of advice. Scientists' apparent frustration about the subsequent delays was expressed in an Editorial in *Nature* under the title 'Why are we waiting?' (Anon, 1976a), and there were strong representations for various details to be settled so that GMAG could begin its work.[46]

As already noted, the DES/MRC and, to a lesser extent, the MAFF/ARC were responsible within Whitehall for articulating the case for following the Williams approach. The broad thrust of the HSE approach, on the other hand, was close to that of the trade unions, though this was not simply a response to their pressure, but importantly a consequence of the formal responsibilities of the HSE.[47] Given this formal responsibility, and public and parliamentary concern, a system of voluntary advice, and temporary voluntary notification pending the introduction of statutory notification, within the framework of the HASAW Act, was probably as much as scientists could reasonably have expected.

An important aspect of trade union action was the need to establish their role in the operation of GMAG, and their objectives, or at least those expressed by ASTMS, had not been so well satisfied as those of the scientists. ASTMS believed that the best safeguard for ensuring that risks were properly assessed, and that precautions were actually observed, lay in worker representation on GMAG and local safety committees. The powers of the HSE should be available to enforce compliance if necessary. These objectives would have been most fully met if GMAG had been constituted as a part of the HSE machinery. However, there was the substantial consolation that the unions would, through their representation on GMAG, have a say in the details of the creation and operation of genetic manipulation policy, and the HSE was still involved in the area, if not as closely as the unions would have wished.[48] The HSE position in the negotiations was certainly compatible with the trade union view, but was additionally motivated by their formal responsibilities under the HASAW Act.

Thus the constitution and remit of GMAG, and its relation to other bodies, was a compromise between these various interests. Arguably, none of these interests accounts for the most interesting and significant aspect of GMAG, the inclusion of those with the specific role of representing the public interest, which would appear to have been a political initiative. These negotiations only settled the framework for decisions about genetic manipulation policy in Britain, the detailed decisions had yet to be taken and put into effect.

In October 1976 the European Science Foundation[49] formally decided to recommend the Williams system as the model for European countries. The stated reasons for preferring the Williams model to that of the NIH were that the former provided a greater degree of flexibility, and had a greater emphasis on well-tried physical containment. At this stage the possibility of a variety of different sets of national guidelines was a cause of concern to the scientific community, and some form of standardisation was clearly desirable. Additionally, current developments in the United States, such as public hearings on genetic manipulation before lay panels and hearings of the US Senate subcommittee on health under Senator Edward Kennedy, which looked as though they would lead to legally enforced regulations, reduced the desirability of the American system.

GMAG's terms of reference
and the representation of interests

GMAG's terms of reference were these:

(1) To advise
  (a) those undertaking activities in genetic manipulation, including activities related to animals and plants, and
  (b) others concerned.
(2) To undertake a continuing assessment of risks and precautions (and in particular of any new methods of physical or biological containment) and of any newly developed techniques for genetic manipulation and to advise on appropriate action.
(3) To maintain appropriate contacts with relevant government departments, the Health and Safety Executive and the Dangerous Pathogens Advisory Group.
(4) To maintain records of containment facilities and of the qualifications of Biological Safety Officers.
(5) To make available advice on general matters connected with the safety of genetic manipulation, including health monitoring and the training of staff.
(6) To submit a report at intervals of not more than a year.
(GMAG, First Report, 1978:1)

Notwithstanding the detailed specification in paragraph 5, the nature of the advice which GMAG would be able to give was left quite open. Indeed, the general tenor of the remit is that GMAG, potentially at least, had considerable latitude in defining its sphere of interest and operation. In line with Brenner's and the Williams Working Party's emphasis on the need for flexibility, however, GMAG was specifically charged in paragraph (2) 'to undertake a continuing assessment of risks and precautions . . .'. Although it was expected that the Williams system of categorisation of experiments would form the basis of its initial work, the scientists had been successful in placing a formal requirement on GMAG to monitor new information on possible risks, and to modify the categorisation system as appropriate.[50]

Beyond this formal remit there was an unwritten but very clear expectation that the Williams report would form the basis of

GMAG's operation, at least in the short term. The next chapter will examine how this worked in practice. For the moment it can be re-emphasised that the 'guide-book' for GMAG had been mainly written by scientists, and that expert scientific knowledge of genetic manipulation was basic to any attempt to assess the risks of experiments.

Besides the chairman, the membership of GMAG included eight people appointed as 'scientific and medical experts', plus four 'to represent the public interest', four 'to represent the interests of employees', and two 'to represent the interests of management'. In addition there were some seven assessors from the HSE, DES, DHSS, the Ministry of Agriculture, Fisheries and Food, and the Scottish Office who took no formal part in decision making. The secretariat of GMAG consisted of two civil servants seconded by the MRC.

As Yoxen (1979a:231) has rightly emphasised, any analysis of the interests represented on GMAG is complex. For example, two of the trade unionists were professors with a direct research interest in genetic manipulation, at least a few of the scientific and medical experts were also trade unionists, and several were concerned with the industrial exploitation of recombinant DNA techniques. Apparently the selection of individuals with potentially cross-cutting interests was the result of a deliberate policy. In looking at this topic in more detail the primary concern is with the interest groupings, but it is unavoidable that particular individuals are mentioned.

Numerically, the largest group on GMAG was the civil servants, the assessors and the secretariat of GMAG. In formal terms the role of the assessors was to provide a watching brief for their respective departments and, where necessary, to provide information from their departments for GMAG. Of the assessors, the most significant was the HSE team,[51] although, as already noted, they kept a relatively low profile in the early days of GMAG. Certainly there was no attempt by the HSE to take the initiative away from GMAG.

The role of the secretariat, who were seconded from the MRC, was considerably more important than it might appear at first sight. They were responsible, together with the chairman, for the detailed organisation of the meetings, and for the day-to-day administration of GMAG and its policies. Although in formal terms the role of the secretariat was the provision of support services, in practice, they were substantially more influential, especially at the beginning, when some other members of the group were uncertain of their precise roles. More concretely, it is clear that the general pattern and orientation of GMAG's detailed work was established by the chairman and secretariat before the first meeting. Given that GMAG was a new departure, the establishment of a *modus operandi* was important.

In effect this identifies one of the main sources of power on GMAG, the access to organisational knowledge and resources. Compared to the situation of many of the members of GMAG who had been only marginally involved in decisions on the development of British policy on genetic manipulation, Dr Vickers, a career civil servant with the MRC, had been associated with the area almost from the outset. He had organised the MRC meeting prior to the establishment of the Ashby Working Party, had been secretary to the Ashby Working Party, was thanked by the Williams Working Party for his assistance, had been closely involved in other MRC actions in the area such as the organisation of the Oxford conference, and the MRC's comments to the Williams Working Party's draft report. Thus he brought a considerable background of knowledge and expertise to GMAG.[52]

The next group in terms of numbers were the scientific and medical experts. Whether or not they had been involved in the development of the policy which had led to the establishment of GMAG,[53] the scientists came to it as members of a community which had successfully channelled the issues surrounding genetic manipulation into the technical consideration of possible hazards and appropriate precautions. As mentioned, the scientists had been chosen on the basis of their knowledge of, and standing in, their particular fields, in order to provide GMAG with a broad base of different kinds of knowledge and expertise.[54]

As far as can be reconstructed from interviews some years after the event, the majority of scientists appointed to GMAG then felt that there were indeed risks attached to the use of genetic manipulation techniques,[55] but that these were not as severe as had initially been thought.[56] Whilst most found the Williams approach more or less acceptable, there was a feeling that the level of precautions was probably higher than required. There was a general recognition that the Williams approach was the price of allowing experimentation to begin.[57]

In general, it seems that the scientists on GMAG saw a major part of their role in terms of the provision of technical scientific knowledge, that is, in terms of a continuation of the approach enshrined in the Williams report. But although the scientists on GMAG were not present to represent an interest, as was the case with the other non-civil service members, the apparent neutrality of their designation as experts did not mean that, in practice, they did not also represent scientific interests to some extent. In formal terms it is of course difficult, if not impossible, to draw a clear line between 'disinterested' or 'objective' information on scientific matters, and articulating scientists' interests. No attempt to analyse these matters will be made, but it can be noted that the margin between the two possible roles seems to have caused some difficulties and self-analysis for one or two of the respondents. For the most part, however, an everyday distinction between the two cases was unproblematically employed on GMAG.

Several participants interviewed stressed that technical information supplied by scientists was potentially open to discussion, clarification and argument in just the same way as any other matter before the Group. In the sense of there being no attempt to hide behind the mysteries of science, there was almost unanimous agreement among those interviewed. However, this should not disguise the fact that the possession of scientific knowledge was an important resource on GMAG, given that in the final analysis experiments and precautions had to be judged on technical grounds.[58]

Turning now to the representatives of the interests of employees, it became clear during interviews that the trade unionists were the only interest group on GMAG to have, and to be recognised as having, a relatively clear set of objectives and a strategy for obtaining them. The objectives derived from the trade union ambitions already noted. Their initial aim on GMAG was to formalise the role of employees' representatives, and more particularly trade union representatives, on the local genetic manipulation safety committees which were established in all centres where experimentation took place.

The four trade unionists on GMAG were a powerful group in terms of the resources and expertise which they brought to the committee. Williamson, of ASTMS, was one of the Scottish scientists referred to in chapter 2, and had been influential in the development of ASTMS policy. Ellwood (Institute of Professional Civil Servants) was a scientist at the Porton Down Microbiological Research Establishment, and his expertise was particularly important for some of GMAG's technical deliberations.[59] Both of these scientists were professors with a research interest in genetic manipulation. Haber was a full-time Divisional Officer of ASTMS, with responsibility for health and safety matters. Owen, the Medical Advisor to the TUC, not only had experience relevant to health and safety issues but, in his previous position as Deputy Director of Medical Services at the HSE, had been a member of the Williams Working Party. Much of the trade union influence on GMAG must be put down to the fact that they were able to field such a well-balanced team which was strong in terms of various kinds of relevant expertise.

The trade unionists normally held their own discussion before each meeting of GMAG, a practice which was frequently remarked upon in interviews, and which caused a degree of concern amongst some members of GMAG. It was felt that the practice signalled an intention to present only a united front, and was perceived by a few as a threat to full and open discussion. This view was usually accompanied by the opinion that the trade union group was responsible for delays in allowing genetic manipulation work to start, through an overemphasis on the formalities of local genetic manipulation safety committee membership.[60]

One significant aspect of discussions on GMAG during its shake-down period concerns a difference in expectations of style and presentation of views. This was primarily a difference between scientists and trade unionists. Several participants explained that the trade unionists tended to treat some discussions as negotiations, and to adopt a bargaining stance, for example, by 'asking for more than they were prepared to settle for'. Whilst this presented no difficulty to the representatives of the interests of management, for example, who were familiar with the rules of the game, it apparently caused discomfort for a few of the scientists to whom this was a novel experience.

There were four people 'appointed to represent the public interest': Jahoda, a social psychologist and member of the Science Policy Research Unit of Sussex University; Lawrie, of the Elizabeth Garrett Anderson Hospital and past President of the Women's Doctors Federation; Maddox, then Director of the Nuffield Foundation, a body which funds research in a number of areas including genetic manipulation, and previously Editor of *Nature*, the premier scientific journal for the biological sciences;[61] and Ravetz, an historian of science and Secretary of the Council for Science and Society. Ravetz also had, in his own words, 'a shaky competence in risk analysis'.

As can be seen, this is a diverse set of backgrounds, although there is a common theme of an interest and concern with the social responsibility of science, Ravetz being the best-known of the four in this respect. As noted, the available evidence suggests that the Secretary of State for Education and Science, Shirley Williams, took a close personal interest in the selection of these public interest representatives. Since the inclusion of such an interest group was a novel departure, and their role unprecedented and not exactly clear, their structural position and role within GMAG will be dealt with separately in chapter 6. For the moment it can be noted that although Maddox in particular had a wide general knowledge of science, the shared pool of detailed technical knowledge of genetic manipulation research was markedly less than in the other groups comprising GMAG.

The interests of management were represented by Gilby, Technical Director of Beecham Pharmaceuticals (UK), and chairman of a technical committee of the Association of British Pharmaceutical Industry; and Windeyer, chairman of the National Radiological Protection Board. The presence of this interest group, in terms of policy decisions rather than general contributions to GMAG discussions, in the first year of GMAG was felt mainly over the issue of the protection of commercial confidentiality which is discussed in chapter 4.

Finally, the chairman of GMAG, Wolstenholme, was Director of the Ciba Foundation. If the intention had been to find a chairman who could hold the ring and unite the potentially divergent interests on GMAG into a body with a common identity and broad consensus of purpose, the choice would seem to have been inspired.

Considerable thought seems to have been given to the composition and membership of GMAG, and care taken to avoid 'committee-sitters' who would take little active part in the proceedings. The structure of the group also appears to reflect a wish to balance the effective power which accrued to scientists by virtue of their technical expertise, through the means of providing a sufficient number of members with counterbalancing interests.

In addition, but with the sole exception of the public interest representatives, each of the groups mentioned had access to both formal and informal mechanisms through which they were able to engage in consultation about the effects of GMAG's actions. Although in one sense all of the groups on GMAG could be said to represent, and be accountable to, a constituency of sorts, in the case of the public interest representatives there was no constituency organisation or mechanism to provide a contact with public opinion or interests at large. Necessarily, this had implications in terms of legitimacy and perceived expertise.

To illustrate this, the civil servants were able to consult with Whitehall, and to act as carriers of opinion from the government machine. This provided a degree of legitimacy and backing to their arguments, and reinforced their particular organisational and bureaucratic expertise. The scientists were able to consult and articulate the opinions of the scientific community and, of course, had the relevant technical expertise. The trade unionists on GMAG similarly had an identifiable constituency which featured elaborate mechanisms for the formulation of views and policies. In addition to their access to relevant technical expertise, the trade union group could also claim to be experts on the opinions of employees in a way that was denied to the public interest representatives in the case of public opinion. The constituency of the representatives of the interests of management is also quite clear, though it is interesting to record that a CBI committee was established at the request of one of the representatives precisely in order to provide a mechanism for consulting the constituency. The public interest representatives, on the other hand, lacked any effective constituency, and could call only on their own individual resources.

4
Operating the regulations

The Genetic Manipulation Advisory Group was originally established to run for two years from December 1976. It was reviewed in December 1978, and again in 1980, and reconstituted in 1984.[1] These three phases in its existence can be characterised as establishment, change, and redefinition of its role, and each will be dealt with in separate sections of this chapter. Each phase coincides with a review of GMAG, its functions and operation, and with a change in chairmanship, as well as a number of changes of membership. Naturally enough, these changes led to differences in the style of the internal relations of the Group, and to some extent to changes of emphasis in its policy. However, there was substantial continuity between the periods into which the analysis is divided, and many of the policy decisions detailed in the following pages represent much more of a development than a switch.

Phase one
Establishment, December 1976–December 1978

Perhaps the main achievement of GMAG's first two years was that it survived. During this period the Group faced, and more or less overcame, issues which at times threatened its external relations with the scientific community, with industrialists, and with the trade unions. The two major issues were those of the assessment of possible hazards and appropriate precautions under the Williams system of categorisation which GMAG operated, and the safeguarding of commercially or industrially valuable information about proposed experiments. Since the eventual change from the Williams system for the categorisation of experiments to a new risk assessment procedure is a major topic it is discussed separately in chapter 5. Commercial confidentiality is dealt with later in this section.

In terms of its internal relationships GMAG was, at its inception, a set of individuals with diverse backgrounds representing potentially divergent interests. During the succeeding two years those interests were extensively discussed, negotiated over, and modified, and GMAG became much more of a group in the sociological sense of the word. The interests did not, of course, disappear, but according to participants, there was a very high measure of agreement on the policy decisions which had been taken and implemented, an emphasis on the generally constructive character of interaction within the Group, and a substantial sense of achievement and even pride.

The establishment of a common identity and a broad consensus on the worth and purpose of the Group was apparently greatly aided by the style of the chairmanship of Wolstenholme. Most of the members interviewed stressed that GMAG was run as a consensus committee; issues were debated, sometimes extensively, until common ground emerged. Little attempt was made to 'force' issues, and only once, it seems, was a policy matter put to a formal vote, and this was on the subject of commercial confidentiality. There were occasional heated exchanges during meetings, and some particular differences within the Group, although there was an underlying unity, aided by the relatively undirected style of the deliberations in which the chairman 'held the ring'.[2]

The internal procedures of GMAG

In outline, the Group met, for the first three years, at monthly intervals to discuss and decide on policy issues, and to categorise proposed experiments. The Williams report provided GMAG with an agenda of areas for policy decisions, and with a system with which to begin the categorisation of experiments.

The general pattern of GMAG's internal procedures was established in the early months of 1977, and particularly important in this respect was the first, full meeting of the Group in January 1977. At this time the newness of the committee meant that many members were still unsure of their roles, and of the possibilities of the Group. Obviously enough, this was the shake-down period in which members had to establish themselves, and the issues. But rather than beginning with generalities, or an extended consideration of the role of the Group, GMAG immediately began discussion of a wide range of specific topics arising from the Williams report, and then turned to the process of categorising the one or two experimental proposals received. That is, there was very little explicit discussion of general policy, but attention was focused rather on 'middle range' policy issues such as medical monitoring and commercial confidentiality, and then on to the details of categorisation. To be sure, there was some limited discussion of the terms of reference of GMAG and its scope of operation. Nevertheless, the Group immediately began to put the Williams report into operation, and did not examine policy alternatives, or set its own independent policy goals.[3]

By all accounts another significant feature of the first meeting was a scientific introduction, given by Richmond, one of the scientists on GMAG, as part of an attempt to provide some basic outline of genetic manipulation techniques for those without technical knowledge. Whilst this was necessary and well intentioned, it was obviously insufficient to give non-specialists anything other than the ability to make but restricted sense of subsequent technical discussions. An unintended consequence was that this provision of information tended to re-emphasise the importance of the possession of technical knowledge for the public interest representatives, and to make clear their disadvantaged position.[4]

One further illustration of GMAG's emergent *modus operandi* drawn from this early meeting was the establishment of the first of a series of subcommittees. These were set up to examine particular issues in greater detail, and to report back to the Group. For example, these subcommittees were concerned with the validation of safe vectors for genetic manipulation experiments, the medical monitoring of those working in the field, and commercial confidentiality. Each of the chairmen of these subcommittees was a full GMAG member, but the overwhelming majority of the subcommittee members were co-opted. Of the fifty-five or so individuals who served on these subcommittees, twelve were GMAG members. Almost all the co-opted subcommittee members were those with scientific or medical expertise, or those from industry concerned with commercial confidentiality. It was open to any GMAG member to attend the meetings of any of the subcommittees.

These aspects of GMAG's procedures illustrate the extent to

which scientific and medical expertise, and technical elements in policy decisions, were the basic currency of the Group's deliberations. Many of the features can easily, and often quite rightly, be defended as necessary, given the general thrust of policy on genetic manipulation, and the establishment of a broad technological paradigm. In other words, these features can be understood as merely symbolic of, and/or as merely reproducing, the centrality of technical knowledge. However, and as outlined below, it would have been within the possibilities of GMAG's terms of reference for it to have attempted to extend its sphere of interest outside immediate technical concerns, although whether scientists would have been prepared to acquiesce to such a move is doubtful. As it happened, no concerted attempt in such a direction was made.

Finally, the few proposals that were submitted to GMAG early in 1977 were quickly categorised under the Williams system. There was very little or no formal discussion of the technical appropriateness of the system, or of general policy on risk assessment. Rather, the Group got its collective head down to the task of implementing the Williams recommendations.[5] They followed the letter of the Williams approach rather strictly, and made no real attempt to follow the spirit of Williams, which recommended a continuing reassessment of the categorisation system.[6] The assessment procedure which emerged was that, while proposals were circulated to all members of GMAG, each of the scientists with appropriate technical knowledge, including two of the trade unionists, was allocated the task of examining a proportion of the proposals in detail, and suggesting a classification. This was then discussed at the next Group meeting.

The scope of GMAG's operations

GMAG's remit was rather wide, and the relations between GMAG and other bodies, particularly the HSE, had been the subject of much discussion and negotiation. At its first meeting the Group was informed that the remit had been deliberately framed to give it 'room to evolve', and that the terms of reference could be changed if it were thought necessary. The relationship between the HSE and GMAG 'had not yet been finally decided'. Even though GMAG was a relatively novel departure designed to deal with a new situation and might therefore require a degree of flexibility, it would appear that this potential freedom to define its sphere of operations goes beyond that necessary, and should also be understood as an invitation to explore the possibilities of action, or advice, on matters outside the arena of the assessment of possible risks and appropriate precautions. In the event very little such exploration took place. A partial explanation for this is that in the early days there was felt to be a strong pressure from scientists to establish and operate a system which would allow experimentation to proceed as rapidly as possible.[7] Subsequent problems with the Williams system occupied GMAG's attention, and by the time the Group was in a position where it might have dealt with wider issues, possibly the desirability of certain forms of genetic experimentation, the scientific consensus was that the possible risks of genetic manipulation had originally been exaggerated, and that such exploration was unnecessary. Even so, and given that GMAG was operating within the technological paradigm to a large extent, it is still somewhat surprising that no concerted attempt was made by the Group to exploit the potential of its terms of reference. A rather circumspect move in the general direction of widening GMAG's sphere of interest was made, mainly by the public interest representatives, towards the end of 1980.

Procedures for the submission of proposals and relations between GMAG and scientists

The Williams Working Party had recommended, *inter alia*, that scientists should first submit experimental proposals to a local genetic manipulation safety committee which would discuss details of the experiment and safety precautions in the light of local knowledge, and agree on a categorisation.[8] The proposal would then be submitted to GMAG who would advise on, or confirm, the level of containment required.

Before GMAG had ever met, scientists at a few laboratories at the forefront of British molecular biology had begun to make tentative arrangements for experiments on the basis of the Williams report. In addition, the MRC Secretariat of GMAG had drafted forms which scientists would use in their submissions to the Group. The information required for the completion of the forms was fairly substantial. It included, for example, the names, qualifications and experience of those directly involved in the experiment; similar details of, and counter-signatures by, the Biological Safety Officer, the Deputy, and the Supervisory Medical Officer; the membership of the safety committee; and a series of details of the experiment itself, covering such aspects as the source of the nucleic acid used, the host-vector system, and the experimental manipulations involved.

Right at the beginning of GMAG's operations there was a brief series of negotiations with some scientists which centred around the issue of what was to count as an experiment. Probably in view of the amount of detail required on the proposal forms, one or two laboratories adopted the position that in sending a proposal to GMAG they would be requesting advice on what was seen by GMAG as a series of experiments rather than a single experiment. The definition of an experiment can be a rather nice conceptual point; if the experiment is repeated with minor changes, is this a separate experiment? And what is to count as a minor change? It seems, however, that the divergence of views between GMAG and these few laboratories went well beyond conceptual niceties. Some of the laboratories appear to have had in mind a kind of blanket approval for all experiments involving specified organisms and procedures, and that this approval might be good for a year, or perhaps longer. In effect, this would have been a licensing system of the type operated by DPAG. GMAG was quite firm, and insisted that there would be no blanket approval, and that experiments would be notified individually.[9]

A related aspect of this mild challenge to the Group was a proposition sent to GMAG by Brenner's laboratory, one of the leading molecular biology laboratories. It set out how the laboratory itself proposed to deal with the internal assessment of experiments and the kind of information which would be sent to GMAG. Most of the members of the Group interpreted this as a 'try-on'. Once again, GMAG was firm in rejecting this attempt to take the initiative and establish a precedent. In general GMAG took a forthright stance and insisted that its formal procedures for the submission of proposals be properly followed. This was particularly so over the question of the composition of the local genetic manipulation committees, an issue of considerable importance to the trade union members of the Group.

The Williams report had established the requirement for safety committees in each laboratory where genetic manipulation research was to be undertaken, and the need for a Biological Safety Officer with 'the necessary training, experience, and authority to enable him to carry out his duties' (Williams, 1976:11). Since the safety committees would be responsible for the first formal consideration of proposals, and would also be in a position to ensure that recommended precautions were, in fact, observed, they would need to be 'properly constituted' and 'representative' (1976:11 and 14), although the precise implications of these important terms were not specified by the Williams Working Party.

From the first meeting it had been made clear that the TUC, and, of course, ASTMS in particular, placed great store on the proper constitution and effective functioning of safety committees. This can be seen as a natural extension of the trade union view that the implementation of effective safety precautions could be best guaranteed if all levels of staff involved, or affected, were properly represented on safety committees as of right. Thus, the trade unionists on GMAG carefully scrutinised incoming proposals and were quick to point out any deficiencies in their composition.[10] A minority, but a substantial minority, felt there was a political motivation behind the trade unionist actions on GMAG with regard to local safety committees. It was suggested that the strong arguments for the principle that employees' representatives on local safety committees should be trade union members was part of a general attempt to establish a trade union role in the operation and implementation of science policy in Britain. These two views of the trade union motivations can, of course, be entirely compatible and complementary. There is no doubt that ASTMS, for example, was indeed keen to establish a role in science policy making, and that a major aspect for this was a desire to ensure the effective implementation of safety precautions in laboratories.

Partly as a result of the trade unionists' arguments, GMAG came to lay strong emphasis on the establishment of properly constituted safety committees. The first of GMAG's many advice notes was devoted to this topic. Thus, in some cases, laboratories were advised that the Group gave its blessing to their proposal, but subject to the establishment of a properly constituted local safety committee, or, in other cases, GMAG withheld comment whilst changes to the safety committee were made. In a very few cases it was necessary for members of the Group to visit laboratories specifically to resolve such problems.[11]

In terms of the operation of the local genetic manipulation safety committees, the available evidence suggests that in the first few years of GMAG, when the hazards of genetic manipulation techniques were generally taken as quite a serious possibility, local safety committees conducted thorough, formal examinations of the proposals before them. During later periods of GMAG's operation evidence indicates that laboratory workers (scientists, technicians and ancillary staff) were more concerned with, for example, the storage of inflammable solvents than with any risks attaching to genetic manipulation experiments. By this time the containment levels for genetic experimentation had been significantly lowered, and the amount of detail required by GMAG on proposal forms greatly reduced. As a consequence, the activities of local safety committees at this time seemed to be merely formal.

In GMAG's relations with scientists, it should be noted that the number of laboratories, and hence of scientists, with whom the Group had contact was relatively low at first, although it was continually rising as additional laboratories became involved in genetic manipulation research. Thus, by the middle of February 1978 GMAG had received 102 proposals from twenty-seven centres (GMAG First Report, 1978:8).

It seems that those senior scientists who had been closely associated with the development of British policy on genetic manipulation had become disappointed with GMAG's early performance. They felt that the Group was being too strict in its interpretation of the Williams report,[12] that it was over-concerned with the letter rather than the spirit of the report, especially on the question of flexibility, and that its procedures were too slow and cumbersome.[13]

It may be that some friction between GMAG and scientists was politically necessary and even beneficial. If genetic manipulation regulations were not seen to produce a few squeals from scientists, then the Group was not doing its job! It was important from the point of view of public opinion, and, in effect, that meant certain members of Parliament, that GMAG should be seen to be having some effect. Be this as it may, it is apparent that the friction became greater than was functionally necessary. GMAG came to be regarded by some scientists as inhibiting the development of British science. Criticism became focused around what was seen as the excessive stringency of the levels of containment required, and also on some of the anomalies in categorisation which resulted from the Williams system for the assessment of precautions.

A number of pieces of evidence can be used to give an indication of the problems which GMAG apparently caused for scientists and an outline of the manner in which they were resolved. It should be noted that the scientists most affected were those with substantial research interests in the area, and who tended to be the more important scientists from prestigious laboratories. As a consequence of this, an important concern was the relative competitive advantage they felt compared with their peer-competitors in the United States.[14] In the rhetoric of the complaints the issue was often cast in terms of the relative disadvantage of British science, but personal reputations were also important.

GMAG's category II containment level, the highest that most experimentalists were able to use because of the shortage of category III and IV containment facilities, was taken to be more stringent than the American NIH equivalent. No direct comparison is possible since the NIH and GMAG systems were based on different approaches, but, overall, it does seem that there was at least some justice in the scientists' complaints in this regard.

In the summer of 1977 the MRC established a Genetic Manipulation Users' Group (GMUG) because they felt that there was a need for a body to represent the interests of the scientists and act as a pressure group.[15] GMUG, under the chairmanship of Brenner, and with Vickers of the Secretariat of GMAG as its secretary, had only two meetings before it was disbanded. Initially, it explored a number of mechanisms for the promotion of research in genetic manipulation such as the sharing of proposed high containment facilities. However, and according to Brenner, a recurrent feature of the meetings was a barrage of complaints about GMAG, especially the anomalies of the Williams categorisation system and the level of containment required. GMUG did propose to GMAG that they should hold a meeting to discuss these issues, but this did not take place. As it happened, other mechanisms were available, and other events would seem to have made such a meeting redundant. Almost all of the members of GMUG were already members of one or other of GMAG's subcommittees, particularly the Safe Vectors Subcommittee. It was from the Safe Vectors Subcommittee that a solution to the problems eventually emerged. Based on an initiative by Brenner in December 1977, the subcommittee developed a new risk assessment scheme which did away with the anomalies of the Williams system, and allowed for the reduction of the level of containment required in some cases.[16]

It seems doubtful that there was a deliberate or conscious strategy of infiltrating GMAG to produce this result. The scientists had been members of GMAG subcommittees for some time, and in any case, the resolution of the friction between GMAG and the scientific community was to the great benefit of both parties; and GMAG made sure that it was thoroughly satisfied with the new scheme before it sanctioned its introduction. Nevertheless, the degree to which scientists were incorporated into the decision making process was increased.

Confidentiality

The possibility that the information requested on GMAG's proposal form might lead to difficulties was realised by the

Group's Secretariat from the beginning, and the topic was addressed at the first meeting. It became immediately apparent that some GMAG members with commercial links, working for, or holding consultancies with, industrial companies, did not want to see any proposals which had commercial implications. This would obviously include proposals from the research departments of industrial concerns. In addition, many of the scientists involved in academic proposals would themselves have commercial associations, and their research might well have industrial implications.[17]

Conscious of its status as an experiment in the operation of science policy, GMAG recognised that its solution to this problem might be regarded as a precedent in the development of a publicly acceptable style of monitoring, and should therefore be given careful consideration. To this end a subcommittee on the confidentiality of proposals was set up to examine the problem and report back to the Group.[18]

Early on in the debate, GMAG took the formal stance that commercial expediency, which might dictate that certain members should not see, or discuss, certain applications, was not a desirable solution, partly because it could breach the principle of the proper scrutiny of all applications. However, the initial message from the subcommittee was that industry simply would not be willing to submit proposals to GMAG if industrial members of the Group were to have any access to them.[19] The first concrete suggestion from the industrialists on the confidentiality subcommittee was that commercially sensitive proposals should be dealt with by a special subcommittee of four GMAG members, chosen for their lack of any commercial or industrial links.[20] The members of this subcommittee would be asked to sign a confidentiality statement which would be binding in law.[21] Together with certain other conditions, this was presented to GMAG as the minimum requirements of industry. Along with it came the implied threat that anything less would mean that the industrial representatives[22] would have to resign from the Group. In general, GMAG was very far from happy with this proposal, and the matter was effectively returned to the confidentiality subcommittee for further consideration.

From the point of view of industry there were two identifiable areas of concern. First, any application for a patent requires that there should not have been any prior disclosure of the details of the process. It was not initially clear whether notification to GMAG fell into this category, or whether it was allowable as privileged communication. Prior disclosure can be verbal as well as written; privileged communication, for example, to certain government departments and Health and Safety Inspectors, is allowed where there are established rules for the maintenance of confidentiality. At first, it seemed that communication to GMAG was indeed privileged in this sense, but the industrialists on the subcommittee felt that GMAG, unlike the HSE for example, was under no obligation to industry to maintain secrecy. At various stages in the first half of 1977, the opinions of government solicitors and learned counsels were sought.[23] At another point it was argued that all communication to GMAG was covered by the Official Secrets Act which members had signed on joining the Group. In this case industry countered that the procedures for the enforcement of this Act were too cumbersome to be useful, and that it would then be up to the government to decide whether to prosecute. Industry expressed the fear that the relatively large number of people on GMAG meant that, were leaks to outsiders to occur, it would be very difficult to trace their origin.

The second concern went beyond that of patent protection and was substantially more important. Whilst patents were undeniably valuable, there are well-tried methods of circumventing the letter of them, and there seems to have been an overemphasis on them in the debate. More important to a commercial concern is the mere knowledge of its working in a particular area, or information on the stage that it has reached in the development of a new product or process. Such knowledge might easily be passed on accidentally, or could be picked up by someone familiar with the area from subtle clues which might be let slip in casual conversation. Thus, it was secrecy about research intentions and progress that was more fundamental to the objections of industry, rather than the more formal arguments concerning patent protection.

In the summer of 1977, the confidentiality subcommittee[24] had a second meeting at which it considered a proposed 'special confidentiality scheme'.[25] The industrialists argued for the formation within the main Group of a mini-GMAG of people without industrial connections. This would avoid any possibility of their inadvertently overhearing, or otherwise gaining any knowledge of, sensitive proposals. The main opposition at the stormy meeting came from the trade union interest group. They were of the opinion that GMAG was no different in kind from such quangos as the National Research and Development Council or the National Enterprise Board, where members, including trade unionists, had access to commercially valuable and confidential information.[26] However, and specifically in recognition of the fears of industrialists,[27] they suggested that a register of interests of GMAG members be set up and that, subject to the chairman's approval, scientists could request that their proposals should not be passed to specified GMAG members.

Following the meeting, a revised version of the special confidentiality scheme was prepared.[28] In essence, this proposed a fixed mini-GMAG of four, which would deal with confidential submissions at the proposer's request. However, when this further version was considered at the next full meeting of GMAG, it was unequivocally rejected by three of the four trade union nominees, who found it unacceptable that no trade unionist should be able to see any of the proposals under this scheme.[29] Because of the strength of the views expressed by the trade unionists, the proposed scheme was not put to the vote,[30] but a temporary alternative agreed. For the time being the chairman would decide, on the basis of a register of interests, who would be excluded from sight of confidential proposal forms, from discussions of them, and from the receipt of related minutes.[31]

Although a temporary compromise had been put together, the underlying opposition of views had not substantially changed. Some of the trade unionists, in particular, appear to have taken the view that the position adopted by industry was, in effect, a negotiating stance which could be modified if the Group took a determined line on the question of confidentiality. An alternative view was that if the Group was so determined they might well find that the multinational industrial concerns would simply export their research efforts to a more favourable regulatory climate.

The issue was allowed to rest for the next few months, but by November 1977 there were several confidential applications potentially ready for GMAG's attention,[32] and the subject was raised again. The outcome of these further discussions was that the temporary compromise, together with a statement of confidentiality to be signed by all members of the Group, would be operated for a further six months, and then reviewed. The particular proposal for the confidentiality statement was, however, thought by some members of the Group to be too inclusive, and it was referred for redrafting.[33] The revised version was circulated to members, and by the time of the December meeting most had signed it, though it still met opposition from the trade unionists, who felt that they would be prevented from having necessary discussions with people outside the Group. Lewin (1977) cited the example of a proposal to clone human insulin genes as a case where the principles of commercial confidentiality and such discussions might come into conflict. Despite considerable argument and discussion, no further compromise could be reached, and for the first time on GMAG a matter was put to a formal vote. The scheme was accepted *nem con*; the trade unionists merely abstained. They did not, however, sign the confidentiality statement.

During the winter of 1977–8, there was a series of contacts between the chairman, Wolstenholme, and the trade unionists, where aspects of the confidentiality statement were 'clarified'. Of particular importance to the trade unionists was the understanding that consequent to the division of the proposal form into two parts, general information and specific technical details, general information, such as the names of the personnel involved and the composition of the local safety committee, could be properly discussed in exactly the same way as with non-confidential proposals. Proposals under the special confidentiality scheme would only be accepted by GMAG if the local genetic manipulation safety committee had had a full discussion of the details.[34] Further, discussion by GMAG members of the specific technical details of a proposal with members of local safety committees would not constitute prior disclosure. With these reassurances three of the four trade union members agreed to sign the confidentiality statement in February 1978.[35]

Following this agreement, the Group began to process applications under the special confidentiality scheme which, although initially a temporary compromise, remained substantially unchanged for the life of the Group. Modifications were proposed at various points, but whenever the issue was raised it rekindled the strong polarisation of opinion, and little progress was achieved.[36]

The problems with, and arguments over, confidentiality were the major divisive issue in the first few years of the Group and severely tested its ability to reach a compromise, if not exactly a consensus. To the extent that the scheme did allow the Group to process applications with as wide a scrutiny as possible, whilst ameliorating the fears of industry, it must be counted as a qualified success.

The debate on GMAG was, of course, symptomatic of more general issues related to the social control of science and technology. From this perspective, Yoxen has argued that the debate 'highlights the real issue which is the need to accommodate patent legislation to public participation, and not the other way round' (Yoxen, 1979a:232). It should be noted that the trade unionists were the only interest group on GMAG able to effectively counter the initial requirements of industry for secrecy and patent protection, and to force the eventual compromise.[37]

In looking more closely at the details of the negotiations, it is also apparent that the appropriate explanation for sets of actions increasingly departs from the general principles which can be construed as the framework for the debate. As already noted, the possible availability of knowledge, in the sense of 'knowing who was doing what', was more significant than patent protection. In the later stages a more subtle set of factors was operating. Formally, industry was concerned with patent protection, somewhat less formally, with the restriction of knowledge about progress and intentions.

Here the possibility of a leak through the trade union organisation was the major early worry.[38] However, interaction and discussion on GMAG and the confidentiality subcommittee brought a certain degree of mutual reassurance to at least some industrialists.[39] What was needed from GMAG was agreement on a set of arrangements of sufficient stringency to maintain confidentiality and give the more liberal industrialists the possibility of convincing their colleagues to give it a chance. Another important factor was the position of industrial research directors *vis-à-vis* their company boards of directors. The former were responsible to their boards for the maintenance of industrial secrecy over research activities. It was stressed on several occasions that the position of research directors demanded that they should not be seen by their boards to risk this secrecy in any way. Thus, towards the end of the negotiations the effective criterion for the acceptability of a confidentiality scheme had moved from patent protection to a scheme which research directors could defend before their boards of directors.

Analogous factors operated in the case of the trade unions. Those involved in the horse trading negotiations on GMAG and the subcommittees seem, like their industrial co-members, to have been convinced of the potential value of GMAG, and did not want to withdraw from it, or otherwise threaten its continued existence, or impair its functions. The trade unionists on GMAG had adherence to the principle of open discussion to justify to their constituency, and that, in the end, they abstained rather than voted against the special confidentiality scheme, would seem to be symbolic of this tension. Subsequently, of course, the justification was eased; once the scheme had been introduced despite their resistance, the new issue was whether to take part in it or not. The 'clarifications' tipped the balance for three of the four trade unionists.

Other policy developments

At the beginning of this chapter it was noted that a number of subcommittees were set up to deal with particular, mainly technical, policy matters. The deliberations of several of these subcommittees were overshadowed, during 1977, by the confidentiality issue and, during 1978, by the development of a new risk assessment scheme, which was initially developed on the subcommittee for the validation of safe vectors.

The subcommittee on genetic manipulation in plants, and a later, related subgroup which considered questions concerning glasshouse containment, were established to deal with the special problems of genetic experimentation in this area. Whilst research using micro-organisms could importantly rely on various physical measures to contain the engineered organisms, one eventual purpose of genetic manipulation in plants would be to release, for example, new or genetically enhanced crops into the environment. Even though this eventuality might be a long way off, experimentation would require that substantial numbers of plants be grown within the foreseeable future. Thus there was, in the first instance, a need to consider what containment measures were appropriate, and feasible, in the case of plants. In the longer term there was a range of potentially difficult issues associated with the release of genetically engineered organisms into the environment.[40]

Although the subcommittee held several meetings during the first two years of GMAG, no policy was implemented by the Group at this stage. In part this was because possible hazard categories and containment facilities could not necessarily be related to those in the Williams report. By the time that the subcommittee produced its first report towards the end of 1978, GMAG was then, in any case, in the process of changing the basis on which it categorised risks and appropriate precautions. One further factor to be noted is that:

> The sub-committee found it difficult to conjecture possible hazards to plant life from genetic manipulations, except those involving existing plant pathogens; it was argued that any new dangerous combination would, in any case, be either self-limiting or detectable at an early stage and could thus be effectively controlled. (GMAG First Report, 1978:27)

Scientists were by now increasingly convinced that the initial fears about genetic manipulation had been greatly exaggerated, and that the possibility of experiments on plants which could not be controlled in a manner similar to other forms of genetic manipulation research was found to be more distant than had at first been thought. Indeed, it is only in the light of this change in the perception of possible hazard that the scientists' difficulty in producing conjectures about them is anything other than startling.[41]

The issues before the subcommittee became somewhat complicated. A technical distinction was made between experiments involving the use and growth of plant cells grown in artificial media which were treated in the same manner as normal genetic manipulation experiments, and experiments which involved the growth of whole plants outside of normal laboratory conditions. This second category was subdivided according to whether any manipulation of a plant pest was involved.[42] If so, the Ministry of Agriculture, Fisheries and Food (MAFF)[43] or the Forestry Commission Secretariat had to be notified at the same time as GMAG, with the former being responsible for issuing the licence without which no experiment could be performed.

Subsequent deliberations by the subcommittee during GMAG's second period led to the publication of an advisory note in January 1980 (GMAG Third Report, 1982) which incorporated these considerations. In essence, laboratory experimental plant cultures were treated in the same way as other experiments; work involving the growth of whole plants was allocated to one or two glasshouse containment categories depending on whether or not a plant pest was involved. The lower of these containment categories, glasshouse containment A, basically required that a separate glasshouse or compartment be used, and that the principles of good glasshouse hygiene be observed. The higher level, glasshouse containment B, was significantly more onerous than the rather limited precautions required by the A category. It required, for example, negative pressure and air filtration, and the sterilisation of clothing, tools and effluent.

This is only an outline of the technical aspects of the subcommittee's work. However, the deliberations of the subcommittee illustrate the complex web of lines of responsibility, departmental interests, and legislative control that were involved. The effective legal framework behind GMAG's activities was the Health and Safety at Work Act, which did not, of course, apply to plants and animals.[44] However, MAFF and the Forestry Commission Secretariat had considerable existing powers available which could be exercised on the advice of GMAG. Thus, the practical relationships concerning plants were analogous to the relationship between the HSE and GMAG, although it was never put on quite such a formal basis. From a formalistic point of view the lines of control seem rather *ad hoc*. To put it another way, GMAG and government departments can here be seen engaged in the process of constructing a system of control out of, and through the adaptation of, existing legislative resources.[45]

The medical monitoring of researchers engaged in genetic manipulation and the mounting of epidemiological studies were part of the Williams remit for GMAG. The Group established a medical monitoring subcommittee to examine the area in detail, and formalised the substance of the Williams recommendations into advice notes on the local safety committees and medical monitoring.[46] The provisions of these notes covered the roles and responsibilities of the Biological Safety Officer and the Supervisory Medical Officer, and such matters as medical examinations, annual health checks, the indefinite storage of blood serum samples, checks to be made in the case of unexplained illnesses, and the keeping of various records and log books.[47]

These provisions, which applied to work in categories II, III and IV, were part of the wider code of practice recommended by the Williams Working Party, and which covered a range of factors concerning the conduct of experiments (GMAG Third Report, 1982:18). The Williams committee had not, however, made any recommendations relevant to the lowest category of work. As a temporary measure, GMAG recommended that such work should be done on the basis of the provisions contained in the Public Health Laboratory Service booklet *The Prevention of Laboratory Acquired Infection* (Collins *et al.*, 1974).

Following the introduction of the new risk assessment procedure for the categorisation of experiments in 1979, and the subsequent lowering of categorisation levels for the great majority of research work, a new containment category, below that of category I and requiring no physical containment, was introduced. This category was known as 'good microbiological practice' after the principles which were to be followed. Basically, and as the name suggests, the procedures and precautions involved were those which should be followed in any microbiology laboratory; for example, no smoking, eating or drinking in the laboratory, the wearing of overalls, and so on.[48] By this time genetic manipulation techniques were thought to hold very little, if any, potential for hazard. However, since an increasingly large proportion of all proposed experiments fell into the lower containment categories, the provisions for monitoring were extended to the lower categories.

The question of epidemiological studies was also affected by the changing scientific consensus on the possible risks of genetic experimentation. Initially, the subcommittee seems to have been in favour of conducting a continuing epidemiological survey; GMAG's second report (GMAG Second Report, 1979:18) noted that such an exercise was to be undertaken by the MRC. However, by the time of the third report there was thought to be no case for an epidemiological survey, although it also pointed out that a retrospective survey could be mounted from the record maintained by the HSE of workers involved in genetic manipulation experiments.

It is an orthodoxy of opinion in the area that no case of untoward effect as a result of genetic manipulation experiments has ever been noted (GMAG Third Report, 1982:9). There is also a body of rather more direct experimental evidence which indicates that harmful genes are vanishingly unlikely to be passed on to humans or to become accidentally incorporated into organisms which might affect humans (see Bennett *et al.*, 1984:ch. 1). Against this it can, of course, be argued that if there were any subtle effects of the use of genetic manipulation techniques or of particular experiments, these might only show up in an epidemiological survey. But, whatever the arguments, it is a testament to the strength of the change in scientific and medical knowledge, that no survey was thought necessary only a few years after the possibilities of hazard were first taken to be worthy of investigation.[49]

The legal and semi-legal framework
of GMAG's activities

It is clear that the relations between GMAG and other bodies were complex and convoluted.[50] Various formalistic niceties surrounded the operation of the Group which was able to give advice to almost anyone on any matters relating to genetic manipulation, but which had no real powers of its own. A series of indirect and informal constraints had been identified to ensure, or in the hope of ensuring, that researchers would be the subject of effective, or at least arguably effective, controls. The major example was, of course, the possibility of withdrawal of research council funding as a sanction to ensure compliance with the interim voluntary notification of experiments to GMAG, and the following of its advice. The possibility of action under the HASAW Act, if there were any accident, and if GMAG's advice had not been followed, had a similar status, and applied to both academic and industrial scientists. In due course, this area was formalised by regulations under the HASAW Act, and a system of statutory notification came into operation in August 1978.[51]

But even then controls did not take a statutory form and depended on indirect and informal constraints. A related informal mechanism was created to cover experimentation involving plants and animals.

One consequence of the lack of any statutory control was that there were no regulations covering the use of the products of genetic manipulation.[52] Once organisms had been transformed by genetic manipulation techniques they were strictly outside the area of the genetic manipulation regulations. Commenting on this, the first report of the Group stated that:

> However, the Health and Safety at Work Act provides the means of protecting workers and the public from hazards arising from the use of the products of genetic manipulation in the same way that it protects them from any other hazard arising from work . . . In particular, in Section 6 of the Health and Safety at Work Act are provisions governing the supply of substances or articles for use at work, laying an obligation on the manufacturer, importer or supplier to ensure, so far as is reasonably practicable, that if they are properly used there will be no untoward hazard. It seems to us that, for the time being at least, here is basis for control of the use of the products of genetic manipulation. (GMAG First Report, 1978:6)

The drawback of this form of 'control' is that it depends on the fear of legal action in the case of actual harm having been caused. In this eventuality, phrases such as 'so far as is reasonably practicable' and 'if they are properly used' offer considerable scope for legal argument.[53] The caveats in the last sentence of the passage quoted above indicate that GMAG was not wholly convinced by its own argument. This impression is reinforced in the second report:

> The Regulations do not include . . . 'use' of the products of genetic manipulation but GMAG has advised HSE that the Regulations should be amended to include it; and in anticipation of such a change the Group has asked for the voluntary submission of brief information, on a simple proforma, of the nature of the work. (GMAG Second Report, 1979:13–14)

Once again, an understanding was promulgated whereby voluntary notification was substituted for a statutory or formal system of control.

In fact, the regulations were not changed. One reason for this may have been the very extensive and lengthy consultation procedures that the HSE was bound to take before making any change. In the third report, the Group reverted to the original position:

> Although work involving the 'use' of a genetically manipulated organism is not covered by the genetic manipulation regulations we consider our present terms of reference sufficiently wide to include this activity in our overall surveillance of the field . . . (GMAG Third Report, 1982:7)

Finally, it should be noted that GMAG's advice was accompanied by a legal disclaimer. The Group were unhappy about this, but they were forced to disclaim responsibility since any negligence on their part would be a link in a chain of legal responsibility which could extend back to Ministers.

GMAG's consideration of its future

At the end of its first report, published in May 1978, the Group reviewed its progress. It was felt that a tolerably satisfactory system had been established, but that there was further work to be done on such matters as health monitoring, confidentiality, hazards to plants and animals, and on experiments involving large amounts of material.[54] The basic debate over the reality of the hazards of genetic manipulation was not settled one way or another. Thus the Group felt that a further term of office was in order. Ministers agreed, and the Group was reappointed for a further two-year period. There were some changes in membership.[55] The chairman, Wolstenholme, retired and was replaced by Sir William Henderson, who had been the Secretary of the Agricultural Research Council at the time of the 'Berg letter', and had been closely involved in the early development of British policy on genetic manipulation. Of the members of GMAG who expressed a willingness to serve another term if asked, only the public interest representative, Ravetz, was not reappointed. There is no firm evidence on where or why this decision was made, but it was suggested in one or two interviews that the explanation was to be found in Ravetz's critical probing of the implicit assumptions on which GMAG, and British policy on genetic manipulation, operated.[56]

During GMAG's first term of office the emphasis was on getting the show going, and then on the details of middle-range policies, almost all of which derived from the Williams Working Party's remit, or from the situation in which the Group found itself. A

101

complex web of tacit understandings and practices grew up which, although they left the Group without any formal powers or sanctions, do seem to have been effective in that scientists and industrialists did, as far as is known, actually follow GMAG's advice. The Group successfully established itself, and was tolerated, if not exactly welcomed, by the researchers it advised.

Phase two
Change, January 1979 to January 1981

A degree of change in the style and direction of GMAG's approach can be seen in the first few months of its second term. To some extent this was part of the general strategy of the new chairman, Henderson, and can be illustrated by his early statement to the Group that he considered the potential benefits of genetic manipulation to be greater than the potential hazards.[57] One implication of this was that GMAG should do nothing that might hinder the development of British scientific and industrial efforts in the area, and should positively encourage them whenever it lay within its powers. Henderson had stood in for the retiring chairman at a meeting of scientists in December 1978 when the new risk assessment scheme had been presented. He had been struck by some of the criticisms of the Group's work which had suggested that its activities were inhibiting the rate of development of research efforts.

The particular style and intentions of Henderson are, of course, only a part of the explanation of the new direction, because a change was, in any case, on the cards.[58] GMAG had completed much of the middle-range policy making which devolved from the Williams report, had established the local safety committee structure in laboratories, and generally created a working system for advice on genetic manipulation. During 1978 there had been increasing pressure on the Group to solve the problems of the Williams system for the classification of experiments and thereby remove the major source of tension between it and the research community. Although the new risk assessment procedure had yet to be implemented, and required a number of policy decisions, its existence took the immediate pressure off the Group. It was received by scientists as a symbolic gesture of intent. Their complaints were in the process of being considered, and action taken.

In outline, GMAG's activities for the remainder of its life took place against the background of an ever-increasing scientific confidence that there were no real hazards attaching to genetic manipulation techniques, and a lessening public anxiety. Along with this, there developed a growing recognition, some would say an orchestration, of the potential scientific, medical and industrial benefits of biotechnology (see, for example, Spinks, 1980). Broadly speaking, GMAG's controls on genetic engineering were progressively relaxed, although from the point of view of a number of scientists this easing of the conditions had not gone far or fast enough. Two important areas of GMAG's policy decisions, which revolved around the categorisation of experiments and the future of the Group, are discussed below. Some of the other items of policy debated in GMAG's second term of office will be dealt with in summary form at the end of this part of the chapter.

Before turning to GMAG's activities the impact of an international conference at Wye College, Kent will be considered. The conference, in April 1979, was held under the auspices of COGENE (The Committee on Genetic Experimentation of the International Council of Scientific Unions) and the Royal Society. It was called to

> review the current status of recombinant DNA research . . . the conjectural hazards, and the regulation of the research through guidelines and legislation. (Morgan and Whelan, 1979:ix)

The conference, which was heralded as the most important on risks and hazards since the Asilomar conference, was widely expected to produce attacks on the principles and practice of the regulation of genetic manipulation techniques. These expectations were amply fulfilled. Thus, in his welcoming address, Dr Michael Stoker of the Imperial Cancer Research Fund Laboratories, London, set the tone with his comment that:

> Despite an orderly start, based on pre-existing safety legislation which gave Britain some initial advantage over other countries, we are now hoist with our own petard. Those responsible are trying to devise a means of escape from the safety net, or should I say prison bars, which prevent us from reacting rapidly to changing ideas evident in the rest of the world. (Stoker, in Morgan and Whelan, 1979:xx)

A number of GMAG members, scientists and non-scientists, attended, and many felt that the aggressive tone of some of the demands for an end to regulation were ill-judged. In interviews, several scientists on GMAG expressed disgust at the way some participants, though not British scientists, misrepresented evidence about risks. It was left to archivist Professor Charles

Weiner, from the Massachusetts Institute of Technology, to point out to the conference that many of those now most keen to play down the potential hazards of genetic engineering were also strong supporters of the original call for a moratorium (Weiner, in Morgan and Whelan, 1979:281–7, *passim*). He discussed the view that only at that point were scientists acting in a scientific manner in wanting to play down the possible hazards. He felt strongly that scientists were acting just as scientifically, and responsibly, when the initial fears were voiced at Asilomar (Weiner, in Morgan and Whelan, 1979:287).

When questioned about his views, Weiner stated:

> It is my impression that expert opinion was sought from the start, and at every stage when there were questions raised there were enquiries made within the relevant community. Recombinant DNA research was a relatively new field for a number of people, and there could be criticisms that the proper people were not invited to Asilomar. I found, however, that experts were continually, and still are, being consulted . . . I think, though, that several of the people who participated at that early stage have stated that, based on the information they had, not to have acted would have been irresponsible, and, secondly, they were talking about the *possibility* of a hazard.
> (Weiner, in Morgan and Whelan, 1979:297, emphasis in original)

As a result, GMAG was well able to discount the propagandist aspects of the conference, and the evidence presented in the scientific papers did not have any direct or immediate effect on the work of the Group. It did, however, contribute to the changing climate of opinion about the likelihood of the hazards of genetic manipulation.

The categorisation of experiments

The development and introduction of the new risk assessment scheme, and the immediate aftermath, are dealt with in chapter 5. The introduction of the new scheme did not just mark a 'one off' change, since the new procedure was more flexible than the Williams scheme which it replaced. The Williams Working Party had promised flexibility, but GMAG had followed the letter rather than the spirit of the Williams report. The Group's categories had remained static and had become out of line with current scientific opinions. By contrast, the details of the new risk assessment procedure meant that it was responsive to changes in the scientific consensus about the likelihood of the risks, albeit indirectly so.

However, the relaxation of precautions for genetic manipulation experiments did not follow automatically, or even necessarily, from the new scheme.

The risk assessment scheme was based on a rough quantitative procedure for calculating the risk of any particular experiment. This required the input of quantitative estimates of certain experimental parameters, and in their proposals scientists could suggest estimates for GMAG to consider. The details were considered by GMAG's Technical Panel,[59] which also helped to educate scientists in the use of the scheme, and to produce uniformity in the suggested estimates of parameters. In practice, actual experiments were categorised by reference to certain paradigm experiments.

The Technical Panel, in effect, set the norms for the quantitative inputs to the new risk assessment scheme, and became one focus of decision making about categorisations. Proposing scientists had the freedom to provide evidence to support their assessment of the various risk analysis parameters, but the freedom was limited by the necessity of securing the agreement of the Technical Panel. It is clear that in the early days there was considerable informal discussion between the two on the technical aspects of experiments and parameters relevant to risk assessment but there is little information as to content. The new scheme could not be treated as a simple calculation which would automatically produce a classification once the relevant numbers had been entered. Like the introduction of any new technique in science, there was a period of discussion and education during which the correct procedures for its use were identified and learned. In this way, the technical norms of the new system were constructed under the significant influence of the Technical Panel.

It should be noted that the characterisation of the new risk assessment scheme as quantitative can involve a subtle mis-emphasis. In practice, the proper application of quantitative estimates was inextricably bound up with, and dependent on, a more analogical process of comparing proposed experiments with the paradigmatic reference experiments. This is not to suggest that the quantitative aspects were unimportant, or that they did not constrain the eventual classification, but the numbers in fact took second place to the technical judgments of the scientists on the Technical Panel.

The Technical Panel was able to exercise some flexibility. GMAG introduced the new risk assessment scheme to run in parallel with the existing Williams system for an experimental period. Scientists could submit their proposals under the old or the new scheme. During the first year of parallel operation, the

Technical Panel examined a number of proposals submitted under the old scheme and, after discussions with the scientists concerned, recategorised them under the new scheme, a process which often resulted in an effective lowering of the containment level required. Thus the new scheme was indirectly responsive to new scientific information about possible hazards.

A point which was seen as important by many of those interviewed was that in a strict sense the new risk assessment scheme was neutral with respect to the absolute level of containment required. The reasoning behind this view runs as follows. From the new scheme, a quantitative expression of the relative risks of various experiments could be obtained. This could be thought of as an interval or ratio scale without any absolute baseline. This then had to be matched up against the four levels of containment, which were also *assumed* to be a kind of interval scale. As discussed later in chapter 5, the levels of containment would be more accurately thought of as an ordinal scale, since the sizes of the intervals were largely unknown. In abstract, the decision to be made was into which containment category a given quantitative risk should be placed. The first part of the process was characterised as scientific or technical; the second part, the appropriate level of containment, contained a much larger element of judgment, and depended to some extent on factors such as 'what was reasonable given the current level of public apprehension'. This was characterised as a social or political decision.[60]

Thus, although there was some room for manoeuvre in the technical scientific area administered by the Technical Panel, there was a significant area of social or political decision making where the views of the whole Group were required. This second focus of decision making was where the non-scientific members of GMAG could potentially have their influence, but it has to be said that here, as elsewhere on GMAG, scientific or medical knowledge was influential, since arguments about the appropriate level of containment were couched in terms of the likelihood of the risks. To put this another way, the separation of the scientific from the social and the technical, whilst formally defensible, takes no account of the unequal distribution of scientific knowledge within the Group. This is not to imply that scientists acted dishonourably, or that they 'hijacked' the decision making; there was, after all, a high degree of individual responsibility shown by scientists. The majority of GMAG members was convinced that relaxation was in order. The point, rather, is that it was the responsible, individual, use of scientific knowledge that characterised the developments, not an effective system of checks and balances.

In February 1979, the new risk assessment scheme was formally introduced for its experimental period, in parallel with the Williams approach to categorisation.[61] At this point no relaxation had taken place, but the majority of the Group were in favour of some move in this direction. As noted earlier, at about this time, and despite some reservations expressed mainly by the trade unionists, the Group also began to explore the introduction of a new category of containment below that of category I, and to be known as 'good microbiological practice' or GMP.

A further input to the issue of categorisation was the problems created by the GMAG-inspired definition of genetic manipulation contained in the HSE regulations requiring compulsory notification of experiments. A particular class of experiments, which involved the rearrangement of the genes of an organism without the addition of any foreign genetic material, was thought by scientists to be without conceivable hazard, and to require no special precautions at all. These self-cloning experiments were, however, covered by the HSE regulations.[62] A major irritant for scientists was that there were traditional, but less efficient, genetic techniques for achieving the same end, which did not involve recombinant DNA techniques, but which were not the subject of advice or control.

GMAG's second report, published in December 1979 but finalised earlier in the year, stated that the definition of genetic manipulation in the regulations was originally interpreted by the Group to exclude self-cloning (GMAG Second Report, 1979:13). Reversing this, GMAG's advice note 8, published in March 1979,[63] informed scientists that GMAG and the HSE had been advised to specifically exempt self-cloning in certain species. GMAG went on to ask that, as a temporary measure, scientists should voluntarily send the Group limited information on any plans for self-cloning experiments.

With the issue of self-cloning satisfactorily dealt with, GMAG turned its attention to further consideration of other matters relating to the categorisation of experiments. The Technical Panel had been asked to consider the possibility of a new category of containment, equivalent to 'good microbiological practice', and produced recommendations for the Group to discuss. There was some difference of opinion, and residual confusion, about whether GMP was to count as a containment category as such, or should be regarded as a recommendation to scientists which covered work outside the formal categorisation system. However, in practical terms GMP was in fact treated as an additional category.

The outcome of the discussions on the Technical Panel, and

within the Group as a whole, was that, in the summer of 1979, GMAG took the view that the period of tandem operation of the new risk assessment scheme and the Williams approach had shown the former to be a success, and that it should take over from the Williams system. In addition, all experiments requiring GMP, all category I experiments, and certain experiments in category II which did not involve expression of gene products could begin as soon as scientists had notified the Group. Previously scientists had been required to await GMAG's advice before starting.

Expression refers to the process of making a genetically manipulated organism produce, or express, the products of any newly inserted genes. An example would be making a bacterium produce human insulin, after the insertion of the relevant genes. The Technical Panel had, in fact, suggested that all category II experiments, including those where expression was involved, should be allowed to begin on notification to GMAG, but the weight of opinion in the Group was that certain of the products that scientists were likely to want expressed in their experiments were biologically highly active, and might conceivably cause hazard, even in minute amounts. For these reasons, GMAG retained its right to examine the details of such experiments before allowing them to proceed. These revised arrangements did not formally come into effect until January 1980. However, during the latter half of 1979, the arrangements were informally phased in, and the old Williams system of categorisation phased out. A further, procedural, change was that the Group's Secretariat dealt with all proposals involving category II, or lower, levels of containment, consulting with members by post if necessary.

In the summer of 1979, having completed its appointed task, and in the light of the fact that some aspects of its consideration of the technical proposals overlapped with the work of the subcommittee for the validation of safe vectors, the Technical Panel recommended that it should be disbanded. Shortly after this, the Safe Vectors Subcommittee recommended, on similar grounds, that it should be abolished. In September, a new Technical Subcommittee was established under the chairmanship of Walker, chairman of the former Technical Panel. The new committee took over the functions of the disbanded ones, and had the general remit of advising the Group on technical matters.

Following the formal introduction of the new risk assessment scheme and the revised notification procedures in January 1980, the volume of work for GMAG was greatly reduced, since the main responsibility for the categorisation of proposals now resided with the local genetic manipulation safety committees.[64]

In January 1980, the NIH published new, relaxed, guidelines; in March, GMAG once again began to review its policies on categorisation and containment. The chairman felt that it was a waste to circulate details of all experiments when most category II and all lower category experiments could begin as soon as GMAG had been notified. It was suggested that the business of checking the proposals that the Group did receive might be handled by a special subcommittee. A further suggestion for discussion was that GMAG consider the possibility that the routine work associated with category II and lower experiments, presently handled by the Secretariat, should be passed to the HSE. A subcommittee representative of all the interest groups on GMAG was set up to conduct a review, and to report its results to the next full meeting of the Group.

The subcommittee meeting was particularly stormy, and failed to produce any agreed recommendations, except that the frequency of the Group's full meetings could be reduced.[65] The arguments concerning categorisation overlapped to some extent with issues related to the future role of the Group, which are discussed in the following section. One view of the policy to be implemented, championed by Walker, was that all experiments of category II or lower should now be formally exempted from notification under the HSE regulations. The current position was that most such experiments could be started immediately on notification; Walker's proposal was to abolish the requirement to notify at all. Local safety committees would, however, continue to consider and categorise experiments, but without any GMAG oversight. Before handing over responsibility to the local safety committees, GMAG would need to clarify and codify its procedures and notes on the risk assessment system.

Walker argued that genetic manipulation techniques were now amongst the most commonly used in molecular biology, that there was no evidence whatsoever of any hazard having been caused to workers or to the general public, and that the present notification system was burdensome to scientists and caused resentment. A subsidiary argument was that the Group should adopt the NIH guidelines. Walker found it a scientific nonsense that Britain was out of step with the rest of the scientific world in not using the NIH approach.[66]

There was strong opposition to this set of arguments, especially from the public interest[67] and trade union representatives. Whilst they agreed that some simplification was in order, they felt that the question of the risks or safety of some category II experiments, in particular, was still open. The difference of opinion over possible hazards was bound up with a dispute over how proposals which GMAG would continue to receive should be handled. This dispute, in turn, devolved on to the composition of the proposed subcommittee which would do the work. The trade unionists found it totally unacceptable that it should be an expert committee, since this could mean that no trade union representative on GMAG would see any category II, or lower, proposals. The trade unionists asked for, and eventually obtained agreement on, a subcommittee composition which included representation from all interest groups on GMAG.[68]

At the April meeting of the Group, the issues which had stalemated the subcommittee were discussed extensively. Although no formal vote was taken, Henderson eventually supported the proposal to advise the HSE to exempt proposals in category II and below, although there was also a substantial opinion that these experiments should continue to be notified, though in a simplified form. In the case of both views, riders were thought to be necessary; certain experiments which involved the expression of potentially harmful products should continue to be notified as before, and safety committees should continue to scrutinise proposals and should submit a list of experiments to GMAG annually.[69]

Amongst those with continued reservation were the trade unionists, and it fell to them to produce a paper outlining their case, for consideration at the next meeting.[70] The paper, discussed at the June 1980 meeting, contained a series of proposals: category I and GMP experiments, and any which involved the expression of products with high biological activity, should continue to be notified as before.[71] If it was found that the arrangements for the exemption of category I and GMP worked well, then exemption of category II could be considered. Agreement was then reached,[72] roughly on the basis of the trade union proposal, with the issue of the category II experiments being passed to the Technical Subcommittee for further consideration.

The final point to be made on the categorisation of experiments is that, throughout the period under discussion, there were a number of more technical decisions made on the accreditation of particular host/vector systems. These decisions were complex, but the general point to be made is that the considerations focused on various biological aspects of the organisms in question. For example, whether particular genes might be passed on to other organisms, or what damage, if any, the products of particular genes might cause. Although this is justified on general biological grounds, the new risk assessment procedure was the proximate cause of this attention to the biological possibilities of hazard. In effect, then, there was by now a much increased emphasis on biological rather than physical containment as a result of the introduction of the new scheme.

Before examining the deliberations over the future role of GMAG it is as well to briefly note the Group's policy with respect to large-scale production involving genetic manipulation. This included any work which involved more than ten litres of cultured organisms. In the early part of 1979 GMAG received applications to move from experimental conditions through the ten-litre barrier. Throughout his chairmanship Henderson was especially keen to ensure that GMAG should not inhibit large-scale production. One problem for industry was that the Group's procedures depended on concrete proposals and inspection. What industry wanted was some form of general advice before they committed themselves to significant investments of capital. This need was met by site visits and informal advice, much of it from Henderson himself, who had many years experience in the large-scale production of foot-and-mouth vaccine.

The future of the Group
and its responsibilities

Henderson had begun his period of office with a reconsideration and subsequent streamlining of the Group's internal procedures. It seems likely that once these processes had been set in motion, and especially after the decisions on the new risk assessment scheme had been taken, attention would then have been turned to the future role of GMAG. From the evidence of his later proposals, it seems likely that Henderson had the eventual goal of bringing the Group to the point where it could be disbanded, and some or all of its functions transferred to the HSE. Control of work in the area by the HSE had, of course, also been policy goals of both the TUC and the CBI since before the establishment of GMAG.

In the event, a consideration of the future of GMAG was precipitated by the election of a Conservative Government with a manifesto commitment to abolish unnecessary quangos. This could, of course, affect the HSE as well as GMAG. In the early summer of 1979, there began a Whitehall review of quangos. It is likely that the advice from the DES to the Secretary of State was that, for the time being at least, GMAG served a useful function and should be retained. The Group had expertise which the HSE lacked, and there would seem to be much to be gained from allowing GMAG to complete its current round of policy decisions and implementations, such as consolidating the new risk assessment scheme and dealing with the issue of large-scale industrial production. In a sense, the existence of the Group was a buffer to any potential criticism about the lack of public accountability of genetic manipulation research. Lastly, the abolition of the Group would not lead to any significant savings since it was cheap to run.

GMAG was to have had a discussion of its future so that it could tender its own advice to Neil MacFarlane, the Under-Secretary of State responsible for the Group. This, however, did not take place, and MacFarlane was briefed personally by Henderson, in the autumn of 1979. Before the briefing took place, the ASTMS members of the Group made their position clear in a paper which was submitted to GMAG but never fully discussed. ASTMS argued that since an increasing emphasis on industrial health and safety issues was to be expected as companies moved towards the industrial exploitation of genetic manipulation, it would be appropriate to transfer GMAG's functions to an advisory body under Section 13 of the Health and Safety at Work Act. It was pointed out that this would also solve the confidentiality issue for industry and secure trade union representation on the new body. However, and despite the opposition of Mark Carlisle, Secretary of State for Education and Science, to both lay and trade union representation on GMAG, the Group was, in fact, one of the first quangos to be assured of a future.[73]

In the spring of 1980 the future role of the Group was again raised by Henderson in the context of the review of procedures for dealing with proposals to be carried out by the representative subcommittee.[74] As already seen, the committee was unable to achieve even a measure of agreement. Henderson had suggested that the routine scrutiny of experimental proposals for category II and below, which was being dealt with by the GMAG Secretariat, should be transferred to the HSE. In the subsequent discussions the various groups outlined their arguments.

The DES was clear that, for the immediate future, GMAG would stay under their auspices, but this did not preclude the transfer of the processing of some lower category work to the HSE. The trade unionists reiterated the longer-term objectives of the TUC, but stated that the piecemeal transfer of responsibilities to the HSE was not acceptable, and that they would prefer the Group to remain under the DES for the time being. The CBI, through their nominees, said they would be pleased to see the HSE take over in the long run, and would be happy with the partial transfer suggested. It was argued that the novel GMAG system had been appropriate in the early days, but that genetic manipulation was now recognised not to be a special case of hazard, and should therefore be dealt with in the same way as any other industrial hazard. Thus, responsibility should be transferred to the HSE and GMAG disbanded. This would solve the problems over confidentiality. Policy advice could be given to the HSE through a scientific advisory committee in exactly the same manner as in other areas.

A number of the scientists on GMAG, and some of those outside the Group, had by this time overcome their earlier fears of control of the area by the HSE. Were this to happen, the HSE would need some form of expert advice, so providing scientists with an avenue of influence. However, there was apparently no substantial positive desire for the wholesale transfer of responsibility to the HSE. It was more important from their point of view that containment levels should be relaxed and the paperwork associated with genetic manipulation controls reduced. Parts of Walker's arguments in connection with exemption have already been mentioned; he also proposed that once this had been accomplished, experiments which required the approval of DPAG or MAFF could be dealt with by them, leaving GMAG with the role of considering the very small number of experiments in categories III and IV, scale up to industrial production, and any new development which might occur. Of course, if Walker's argument that the NIH guidelines should be adopted were accepted, then there was very little reason at all for the continued existence of GMAG.

The public interest representatives were not in favour of the abolition of GMAG since they considered that it still had important functions to perform. They were, however, willing to consider the transfer of the routine work of processing proposals in the lower containment levels to the HSE. The most important objection to such a transfer came from the HSE itself. It was explained that their present financial circumstances would prevent them from employing and training the staff necessary to deal with added work. However, they were perfectly willing to consider taking over the responsibility for the Group in the longer run. Reading between the lines a little, it seems that the proposed piecemeal transfer would do nothing for the HSE except increase their costs somewhat.

The possibility of any such transfer was effectively blocked, and there was no other immediate outcome of the discussions about the future of the Group. The next move was a letter sent to the chairman in August 1980 by Professor Subak-Sharpe,[75] asking that the Group formally consider its long-term future. Subak-Sharpe argued that recent scientific evidence showed that the earlier fears which had led to the establishment of GMAG had been misplaced, and that there were no hazards inherent in genetic manipulation techniques. No case of harm had ever been found. GMAG had been needed in the early days to give advice to the scientific community, and had, on the whole, done its job well. However, scientists were now thoroughly familiar with the area and GMAG no longer had a monopoly on expertise; it was therefore without purpose. There were a few areas which required tidying up, for example, large-scale production, and some aspects of the risk assessment scheme concerning biologically active molecules, and these should be completed. When this was done control should be transferred to the HSE and GMAG replaced by a committee of scientific experts who could provide technical advice as necessary.

Subak-Sharpe's letter was not debated by the Group until the December 1980 meeting. Before discussing the outcome, it should be noted that the middle-term future of the Group had already been settled, since in the autumn members had been asked to indicate whether they were prepared to serve another term of office. Between writing his letter and its consideration by GMAG, Subak-Sharpe had written to the DES declining a further period of office, and arguing that the Minister should consider abolishing the Group in December 1980. There is some evidence that the question of sponsorship of the Group had also been under consideration since the initial decision of the then new Secretary of State that it should remain with the DES for the time being. Whether there had been a re-run of the strained negotiations in Whitehall, which had characterised the period immediately prior to the establishment of GMAG, is not certain. What does seem likely is that GMAG was to be allowed to complete its current phase of work, especially that involving large-scale production. It was apparently felt, in certain quarters, that it was necessary to allow GMAG to establish a procedure for dealing with large-scale production, as a means of protecting industry from any system that the HSE might devise. The question of an appropriate home for the Group, or of an alternative body to provide advice, could then be settled later, and in the light of the kind of advice that was then needed.

Subak-Sharpe's arguments produced a good deal of heated discussion. Members largely felt that the scientific evidence was not as unequivocal as was suggested, that there were still important issues related to gene expression, and that there was still an important role for the Group. In addition, there was a not inconsiderable amount of work to be done on large-scale production, and a continuing need to assess any new techniques which became available. In the end, only one other member of the Group supported Subak-Sharpe in his call for the abolition of GMAG.

An interesting point about the arguments over Subak-Sharpe's proposal was the emergence of a series of new areas of policy which the Group should consider. These included the ecological and environmental effects of the products of genetic manipulation, for example, the effects of new kinds of plants, or of bacteria designed to deal with the problem of oil pollution, and wider ethical issues such as those associated with gene replacement therapy in humans. Accounts of the discussions which led to these suggestions vary. According to some, these proposals commanded broad agreement; others held that there was little disagreement, but small real consideration of the matter. Thus there was a difference of opinion on the status of these proposals, and they have been represented as examples of what GMAG might do, and which just happened to have been thrown into the general discussion, and also as an agenda for future policy decisions.[76] Most of the proposals appear to have come from the public interest representation on the Group.

But whatever the perceptions of the status of the proposals, it was not until December 1981 that there was any discussion in the Group of issues which could be said to relate to the proposals. Then, GMAG's opinion was that the environmental implications of the use of genetically manipulated organisms and plants were too far off to require consideration of guidelines for their use. At the same time, the increasing international concern over the moral and ethical implications of genetic manipulation applied to humans was noted, and it was recommended to the MRC that they review such matters.

Finally, two other aspects of policy from GMAG's second term of office should be noted. They are symbolic of the Group's accommodation to the changing climate of opinion over the hazards of genetic manipulation techniques, and the pressures from scientists. The first concerns the matter of research relevant to risk assessment procedures. The new risk assessment scheme was formally predicated on the idea that quantitative estimates of various parameters could be obtained, and that these would be experimentally based. Thus, one might have expected that some research would be undertaken to provide the appropriate evidence.[77] Indeed, when the then Secretary of State for

Education and Science, Shirley Williams, was interviewed for the television programme 'Weekend World' in November 1978 about the impending introduction of the new scheme, she explained that she was prepared to consider the funding of experiments on risk assessment to the tune of 'tens of millions of pounds'.[78] Again, when the Technical Panel began to administer the new scheme in the early part of 1979, part of its remit was to see whether there was any research which needed to be done. However, the approach that was taken in practice was to use existing knowledge about, for example, pathogenicity and damage, to provide some empirical backing for the quantitative estimates.

In fact, very little risk assessment research was done in this country. Bemoaning this, Maddox, in an unsigned editorial in *Nature*, commented that:

> It is true that people have fed each other quantities of *E. coli* from time to time, and there have been a few half-hearted attempts to see whether genes incorporated into bacteria can find their way into the somatics [sic] cells of laboratory animals. Few opportunities have been seized for looking into the ways in which bacteria carrying foreign genes might actually damage their adventitious hosts. The customary explanation, during most of the past five years, is that the people most competent to carry out such experiments have been too preoccupied with exploiting the new techniques whose safety was called into question. With a few conspicuous (and honourable) exceptions, those with the most vivid interest in demonstrating that their techniques are as safe as they now appear have been content with arm-waving. (Anon, 1980b:474)

Now, the criticism is addressed to the international scientific community, rather than to Britain in particular. However, the fact that scientists were able to negotiate the reduction of precautions on the basis of limited research based evidence is a testament to the power and influence of their expertise.

The second point relates to the scientific merit clause in the recommendations of the Williams Working Party on the consideration of proposals to conduct genetic manipulation experiments. Their report suggested that discussions should take account of the 'scientific merits and of potential hazards' (Williams, 1976:14). As already noted, this seems to have been borrowed from an analogous clause in the report of the Godber Working Party on the control of research involving dangerous pathogens. Here one might expect that there should be good reasons for wanting to perform such experiments. In 1976, such good reasons could be thought to be necessary in the case of genetic manipulation as well.

This clause found its way into GMAG's advice note 7. It was also part of the notes on the completion of the proposal form, in the section dealing with the consideration of proposals by local safety committees. This matter was raised in the spring of 1979 and led to spirited exchanges on GMAG. Subak-Sharpe, who raised the matter, argued that it was no part of the work of GMAG or a safety committee to comment on the scientific merit of experiments or the research programmes of which they were a part. This was a scientific issue which could only be assessed through the normal process of peer review by other scientists. Local safety committees might contain non-scientists, and scientists from other disciplines who were not professionally competent to comment. In essence, then, the issue was presented as one of the professional control of science, and non-specialists did not have any part to play in this. The question of any hazards of any experiment was to be sharply distinguished from its scientific merit.

The GMAG advice note 7, which contained the offending clause, had been written by the two ASTMS members of GMAG, and on the basis of the clause in the Williams report. Their defence was that the local safety committee should not be prevented from engaging in such discussions which could have important safety implications. Professional control should not be allowed to interfere with full discussions of safety. In the end, a form of words was found which could be agreed by both parties: 'the safety committee may wish to consider the scientific content of each proposal . . .' (GMAG Second Report, 1979:32). Once again, the episode is symbolic of the special status of scientific knowledge.

To summarise, the second term of office was an active period for the Group which was importantly characterised by the accommodation of genetic manipulation controls towards the scientists' current conceptions of the likelihood of the risks, and by a series of considerations of its role and future. After the crucial introduction of the new risk assessment scheme, scientists were able to introduce progressive relaxation, with the trade unionists acting as the only effective counterbalance. In this context, and although they do not figure as a significant force in terms of the arguments outlined, the public interest representatives acted in some sense as honest brokers between the differing factions. Thus, their agreement to various policy decisions may have been more important than a descriptive account suggests. The only point at which wide social and ethical issues were raised stands out as an anomaly, and was of little consequence to GMAG's policy. Finally, it should be noted that throughout the second term there was an increasing need to consider the regulations for large-scale production, even though the Group was surprised at the small number of applications it received. This work was in train at the end of the second term, and was completed in the third.

Phase three
Redefinition of role, January 1981 to February 1984

There is much less to discuss about GMAG's third term of office, mainly because the work of the Group was significantly reduced, and it now met infrequently.[79] The main areas of policy during this time were the development of a system for the notification of large-scale production to replace the earlier version which had been published at the end of 1979, and continuing discussions, largely outside the Group, about its future.

Sir William Henderson resigned from the Group at the end of 1980 on taking up a position on the Science Council of Celltech, a company formed to develop commercially some of the findings of British research in genetic manipulation. Obviously, this would have been incompatible with his continued chairmanship of GMAG. The new chairman was Sir Robert Williams, formerly Director of the Public Health Laboratory Service, and chairman of the Williams Working Party.

Large-scale production

At the first meeting of the year in April 1981, GMAG had an extended discussion on possible procedures for dealing with proposals for large-scale work. It was accepted that there would have to be a strong emphasis on biological containment since high levels of physical containment would be difficult to guarantee in practice. Thus the assessment of proposals would be importantly concerned with the inherent safety of the organism in question.

An *ad hoc* working party under the chairmanship of Walker was set up to examine procedures in more detail. The membership included those from each of the interest groups on GMAG. The working party met a group of industrialists under the auspices of the CBI, and produced a draft set of proposals for the June meeting of GMAG. This contained two important principles. First, the respective roles of GMAG and the HSE were to be clearly differentiated. Although this was not fully clarified until later in the year, it meant that GMAG would be concerned with the biological aspects of proposals (i.e. the particular characteristics of the micro-organism to be used), while the HSE would be responsible for inspection of the industrial plant, as in other areas of biochemical production. Second, different notification procedures would apply according to whether the organisms and manipulations involved were those which were presently exempted from notification as experiments, or fell into one of the containment categories for experimental work.

There were, however, a number of points of disagreement. One of these was the 'ten-litre barrier', and was the brunt of criticism from the CBI. It was argued that, in effect, the existence of the barrier constituted an additional category. The CBI challenged GMAG to justify this. If an organism was safe as far as experiments were concerned, then it was also safe in larger quantities. If an organism or manipulation had been exempted from notification for experimental work, then industry should be allowed to go ahead with large-scale production, subject only to the safety standards that would apply to any industrial plant, irrespective of whether genetic manipulation had been involved. Thus GMAG began to consider specifying precisely what risks were involved in larger-scale production. The other substantial point of disagreement concerned notification. Industry wanted something approximating the arrangements for experimental work which would provide exemption from notification for certain classes of processes. The public interest representatives and the trade unionists, however, insisted that all proposals for large-scale work should be notified.

In September 1981, GMAG had a further discussion on a revised draft set of proposals. There seems to have been little disagreement, and, except for a number of small technical changes, the proposal was eventually published as advice note 12 in February 1982 (see GMAG Third Report, 1982:46). Basically, there were two procedures for large-scale production. Work could begin immediately on notification to GMAG where large-scale work involved the use of self-cloning, other categories of work which were exempt when considered as experiments or the equivalent of good microbiological practice. All other large-scale work which, if carried out as an experiment, would be classified in categories I–IV, had to be notified to the Group, who would be expected to give their advice within twenty-eight days. GMAG would be concerned only with the biological aspects of the large-scale work; all else would be dealt with by the HSE in the usual manner. Thus industry achieved a large measure of its requirements.

During 1981 and the first part of 1982 the Group attended to a number of other policy issues. Following a discussion paper on the possible environmental issues prepared by Ellwood, a scientist and one of the trade union members of GMAG, it was agreed by the Group that there was no present need for guidelines for genetic manipulation in plants.[80] Increased international concern over the moral and ethical implications of the possible application of 'genetic engineering' techniques to humans was also noted, but the Group did not take any immediate action other than to suggest that the MRC might review these matters. This is, at first sight, a surprising response. The formal rationale was that a good deal of the public concern focused on, or overlapped with, issues such as *in vitro* fertilisation, which did not involve genetic manipulation as such, and which was under consideration by the Warnock Committee (see Warnock, 1984). However, this still left a residuum of issues such as gene replacement therapy and, for some members of GMAG at least, this was not the end of the matter. The issue resurfaced later, as did the topic of environmental implications.

Other areas considered by GMAG, in 1982, included a proposed European Recommendation on genetic manipulation, and the American proposal that the NIH guidelines should be substantially relaxed. In its final form the non-binding European Recommendation did little more than call for the notification of experiments to a competent national authority.[81] The proposed revisions to the NIH guidelines, which would have effectively abolished the remaining controls, were modified before acceptance, and did not lead to any significant difference between the stringency of the NIH and GMAG procedures. Thus neither of these issues had any real effect on the work of the Group. Some outstanding matters concerning the containment of modified viruses, a topic on which GMAG had in the past found it difficult to give clear advice, were passed to the Technical Subcommittee for consideration. This left just the question of whether GMAG was still required.

The future of GMAG had been the subject of considerable discussion during its second period of office, and, as noted in the previous section, there was an emerging consensus that the HSE should eventually take over GMAG's functions. An early transfer had been rejected mainly because the HSE did not feel able to take on the task. Less influential reasons were the reservations of the public interest representatives, for it seemed unlikely that there would be such representation on any new body, and the fact that there were several outstanding issues, such as large-scale production, for the Group to settle.

In this third period of office, and when GMAG's third report was in the process of production during the early part of 1982, there was sharp disagreement over a section dealing with the possible future role of GMAG. The draft suggested that consideration of human gene replacement therapy was not within the remit of the Group, and that, in any case, it was not properly constituted to deal with the issue. The trade unionists took strong objection to this, arguing that it certainly was within the remit, and that GMAG could easily co-opt any additional expertise required, just as it had done on many other issues in the past. GMAG, in fact, had a better claim to look at such questions than many other bodies, it was argued, since it had the confidence of the scientific community, and contained representation of a number of different interest groups.[82] It seems doubtful that these objections caused any real change in official policy over the future of GMAG. The offending section of the draft was omitted from the third report which merely stated that the majority view of members was that it should keep the field of genetic manipulation under review, at least for the foreseeable future.

Despite this statement there were already some moves towards a transfer of GMAG's functions to the HSE. During the latter part of 1981, and through the first half of 1982, there was a series of discussions between the DES and the HSE over the future sponsorship of GMAG. The Group was not involved in these, although they were informed that they would be consulted and their opinion taken before any final decision was made. The discussions involved some relatively junior-level officials, suggesting that matters of detail were on the agenda. In turn, this implies that agreement on the principles of a transfer had already been reached and that only the final details were being sorted out.[83]

The results of this lengthy exercise[84] were presented to the Group as the 'Proposed Reconstitution of GMAG' at its meeting in November 1982. The various government departments consulted, and the HSE, were all in favour of the creation of a new body under the HSE, i.e. the consensus of officials was that there was no objection in principle. The body was later designated the Advisory Committee on Genetic Manipulation (ACGM). Following GMAG's comments, the proposal was to go to Ministers for approval, and then through the HSE's consultation procedure. The tone of the document, and of the discussions in the Group, was in terms of the continuity of regulation. The functions of GMAG would be transferred to the new body, but this was not intended to lead to any significant changes in style, or in the regulatory environment.

The Group had a wide-ranging discussion. There was consensus on the need for a scientifically literate secretariat for the ACGM; that the new body, like GMAG, should have the freedom to act on its own initiative if necessary; that the ACGM should be constituted under Section 13 of the Health and Safety at Work Act, thus ensuring trade union and employer representation; and that the government should be presented with a stronger case for the continuing supervision of genetic manipulation, since many scientists felt, and might argue, that there was no further need.

Disagreements revolved around the size of the new committee, and its remit. Broadly speaking, it seems that scientists wanted a small expert committee which, like GMAG, would co-opt additional expertise as required. The trade unionists, in particular, favoured a larger membership, and argued strongly that its terms of reference should be sufficiently wide for it to consider ethical issues in addition to health and safety matters. On the latter point, the trade unionists were in the minority. The need for advice on ethical matters, indeed, the urgent need, was recognised, but it was felt that the HSE was not a suitable organisation to undertake this responsibility.

Turning to the influences behind this proposed change, it has already been noted that the trade unions had argued for the HSE as the sponsoring agency since the time of the Williams Working Party, and the CBI had long been pressing a similar case. With the settlement of the outstanding policy issues on GMAG during 1982, and the routinisation of much of the Group's work, the earlier hesitancy of the HSE to accept responsibility had largely evaporated. Also, the HSE already had the Advisory Committee on Dangerous Pathogens (the reconstituted DPAG) under its wing. Most scientists probably felt that there was now little need for the regulation of genetic manipulation; but again most, and particularly the senior scientific establishment and the MRC, had lost their earlier fears that the HSE lacked the expertise to do the job. A more structural reason for the change was the expectation that regulation would increasingly be of industrial applications. As far as can be established it was accepted in Whitehall and government that there was still sufficient public concern to justify the continuation of some form of regulation. Or, perhaps more accurately, that sufficient public and parliamentary concern would be generated by the complete abolition of regulation to justify its continuation.

There had also been some behind-the-scenes activity concerning the future of the Group. Following its arguments earlier in 1982 for the broadening of GMAG's remit, ASTMS had submitted a memorandum to the DES restating this point, stressing the need for continued supervision of the area, and requesting a meeting with the Minister responsible. The union and its parliamentary group of Mps met William Shelton, Under-Secretary of State for Education, in October 1982, just before the publication of GMAG's third report, and before the meeting of the Group at which it discussed its future. ASTMS were reassured by Shelton that GMAG, or at least its regulatory functions, would continue.

The structure of HSE advisory committees makes no formal provision for the inclusion of representatives of the public interest. This was taken by most of those interviewed to automatically imply that there could therefore be no public interest representatives on the new body. However, it was once or twice suggested that this was not a necessary conclusion, and that had there been sufficient pressure,[85] or sufficient political will, some form of administrative arrangement could have been constructed. In the event neither of these conditions obtained, and there was no discernible public pressure to retain the role of public interest representatives. In the early days of GMAG the incumbents had been enthusiastic in representing the public interest, but that drive had faded with the increasing belief that there were no real dangers, and in the face of the difficulty of performing the role. It had never been easy to recruit, and it had become increasingly difficult; who should one consult, for example? No doubt solutions to such problems could be found, but the Group had managed with a gradual reduction in the attendance of public interest representatives.[86] It seems that there were no arguments on GMAG against the abolition of the role.[87]

It had been expected that consideration of the proposed reconstitution of GMAG would lead to the Group being dissolved in mid-1983, and no further meeting was planned at this stage. It was, however, not until January 1984 that the final meeting took place. The long delay was explained as resulting from the time taken before all the responses to the HSE consultation had been received. The Report to Ministers and the Health and Safety Executive by officials of the DES and HSE clearly came down in favour of the transfer to the HSE. It gave details of the proposed remit, scope of operations, and categories of membership. Thus possibilities of abolition without replacement and reconstitution with responsibility for some ethical issues were not considered.[88] The consultative document had been issued in April 1983 but responses had continued to arrive until August. It appears that there had been a good deal of lobbying, as at previous points in the history of GMAG. Some of this had come from the trade unions on the matter of the consideration of moral and ethical issues. Apparently the proposal went as far as the Cabinet Office, and almost certainly, therefore, to the Prime Minister.

Although GMAG was to be disbanded, the watchword of 'continuity' had substantive implications. At this final meeting the Group considered three important issues, two of which, in effect, set the agenda for the early deliberations of the ACGM. First, and partly in response to a statement from the HSE that the new committee would not deal with moral and ethical issues, the Group noted that the application of genetic manipulation techniques to humans was now a real possibility.[89] There was, therefore, an urgent need for guidance, and since the Warnock Committee would not cover such topics as gene replacement therapy, the Minister of Health should be advised to take up the matter.[90] Second, the issue of environmental impact of the release of genetically manipulated organisms was now also urgent. Substantial concern had been expressed about this in the United States, and some institutions in Britain were approaching the stage of field trials. In addition to strictly environmental questions, the potentially great economic benefits meant that the preparation of guidelines should not be allowed to cause any delays.[91] The ACGM was therefore advised to consider the introduction of guidelines for the release of genetically manipulated plants and other organisms. Third, and a point raised by Haber, one of the trade union representatives, was that there had been a number of important scientific developments relating to what are more popularly referred to as 'cancer genes'. Some scientists had suggested that one implication of the developments was that scientists' earlier confidence that 'cancer genes' did not represent a real hazard was misplaced, or at least in need of reappraisal in the light of the new evidence.[92] Hence the ACGM was recommended to review existing GMAG advice on genetic manipulation in viruses.

Thus, GMAG passed on its responsibilities to the ACGM. The Group was dissolved at the end of February 1984, and the new Committee took over the following day. Continuity was also ensured through the membership. The ACGM was chaired by Sir Robert Williams, the former chairman of GMAG from 1981 to 1984 and of the Williams Working Party, and was composed of eighteen members, eleven of whom were members of GMAG or its subcommittees.

5
Reconceptualising hazards and risks

The 'Berg letter', which first brought scientists' attention to the potential hazards of recombinant DNA research, made it quite clear that the risks involved were conjectural. 'Although such experiments are likely to facilitate the solution of important theoretical and practical biological problems,' it said, 'they would also result in the creation of novel types of infectious DNA elements whose biological properties cannot be completely predicted in advance' (Berg et al., 1974). Some five years later, as was manifested at the Wye Conference in Kent, the attitudes of many scientists had undergone an about-turn. By 1979, scientists were saying that rather than hazardous outcomes being either a possibility or a likelihood, all that could be agreed on with certainty was that the risks could not be completely ruled out (see Morgan and Whelan, 1979, *passim*). As was noted earlier, only isolated attempts were made to experimentally assess the degree of risk associated with recombinant DNA techniques.

Clearly, this poses a serious problem if judgments are to be made about the relative risks of different kinds of experiments, especially when, as Yoxen (1979a:235) pointed out, this is 'compounded by the fact that an institution has already been created' for just this purpose. Wynne (1980:186) has argued that 'Since we cannot in any significant sense assess the technology itself for its full "factual" impact, we have to assess the *institutions* which appear to control the technology'. As part of this process, the following chapter will explore both the social and the cognitive

126

nature of the changes which resulted in GMAG's introduction of the new risk assessment scheme to replace its initial reliance on the Williams proposals. But first, a brief account of some aspects of the scientific background is in order.

Because scientists themselves have continued to argue about what the real causes were for beliefs about the possibility of hazards arising from the use of genetic manipulation techniques, and about whether the call for a moratorium was justified in terms of what was known at the time, it is not easy to provide a comprehensive account of the reasons for the changes in their attitudes. The arguments have devolved on to, for example, questions of whether there was a failure to integrate knowledge about the new techniques with that concerning the pathogenicity and ecology of micro-organisms. The reconceptualisation of the possibility of hazard has gone hand-in-hand with a reassessment of the validity of the basis of the initial concerns.[1] In addition, the relative lack of scientific controversy about the changing perceptions makes it difficult to recover the implicit assumptions.

Historically, concern was first expressed by scientists working with cancer-inducing viruses (particularly a tumour-causing virus, SV40), and using early genetic manipulation techniques. With the development of much more powerful experimental means of fragmenting genes, and recombining them in many, sometimes random, ways, more general fears were aroused. It was thought possible that such experiments might lead to an increase in the virulence of pathogenic organisms, or even to harmful new forms. There were other areas of concern, but the important ones were the possible introduction of 'cancer-inducing genes' into organisms, and the production of unintended recombinants with unpredictably hazardous properties.[2]

By 1978 most scientists were convinced that the possibility of any hazardous outcome had been either greatly exaggerated, or completely misjudged. During the intervening period molecular biological knowledge about the structure and organisation of genes within the genome had significantly increased – importantly as a result of the use of the new genetic manipulation techniques. Rather than the genome being seen as 'just a collection of genes', it was now regarded as having a complex internal organisation in which, for example, the operation of one gene could be affected by other, physically distant, genes. An implication of this was that one could not simply add a 'cancer gene' or a 'virulence gene' to an organism, and expect that section of DNA to be expressed and have an effect. It had been found a difficult task to induce some incorporated genes to function at all, since this depended on a complex set of interactions with other parts of the genome. Thus 'virulence' and 'pathogenicity' were not the functions or properties of specific genes alone, but of the genome as a whole. Of course, it was still conceivable that a new or dangerous organism might still be produced completely by chance. This possibility can never be ruled out, but it was now thought to be vanishingly small. It was argued that if there were any real likelihood of producing a hazard, nature, through evolution, would have done so already.

A component of the early concerns had been, in effect, that hazardous genes might be incorporated into parts of genomes to which they would not normally have access, and could there cause havoc with the genetic machinery, or lead to some harmful effect. The increased knowledge of the molecular biology of the gene showed that the cutting and splicing of genes, the rearranging of them within the genome, and the incorporation of genetic material from other organisms and from other species, were all much more common in nature than had been thought. Genetic manipulation experiments did not differ as much as had been first envisaged from the natural processes they mimicked. Brenner (1978) explained the situation quite forcefully:

> What a genetic manipulator does is to do things *in vitro*. In theory, if you could do genetic manipulation by avoiding the use of test-tubes then in fact you would be doing something that was a natural mechanism, and so in theory you might argue that you could escape from the GMAG regulations . . . Work on recombinant influenza viruses, work on recombinant bacteria . . . goes on, but doing the same work – indeed, the identical experiment could be formulated – with a restriction enzyme would put you under the GMAG regulations.

The increased knowledge, then, indicated that the possibilities for natural genetic exchange, in some cases across barriers once assumed to be largely inviolable, was much greater than had been thought, and that the chances of such exchanges leading to 'successful' recombination in the sense of producing a hazardous outcome were very small. A further corollary to be drawn was that the use of fragmented sections of the DNA of, say, smallpox virus cloned in a bacterium such as *E. coli* was safer than handling the virus itself. Thus, no genetic manipulation experiment should require precautions greater than those needed for handling the pathogen involved.

Before turning to the effects of these changes on the categorisation activities of GMAG, one additional aspect should be discussed. This concerns the possible hazards of cancer-inducing genes. In fact, the evidence and arguments involved extend beyond the period under review in this chapter; however, they deserve mention because they provide a different perspective on some aspects of the reconceptualisation noted above, and because this is one of the very few cases of scientists arguing that tighter restrictions might now be required.

In 1975 the American Recombinant DNA Advisory Committee asked for experiments to be performed to assess the possibility that segments of the DNA of viruses which had been cloned into *E. coli* could later transfer out of the bacterium and gain access to the tissues of the host animal in whose intestine it resided. The DNA fragments used were of the tumour-inducing polyoma virus, mice being the host animals (see Krimsky, 1982:244–63). At the Wye conference, those carrying out the experiments stated that 'Our task was to evaluate whether recombinants containing a viral DNA segment could transfer out of *E. coli*. The answer to that question is an emphatic "No!"' (Martin, in Morgan and Whelan, 1979:209). The results of these experiments were an important plank in the arguments for the view that genetic manipulation posed little risk; the possibly dangerous DNA was safely contained within the *E. coli*. Other conclusions supported by this evidence included the view that there was a substantial safety factor in handling fragments of DNA rather than the whole genome.

These conclusions had been questioned at the time (see Krimsky, 1982:244–63 for details), but the challenge had been set aside by scientists and the RAC. In 1983 articles were published by Bartels and others arguing against this reassuring picture (Bartels, 1983; Bartels *et al.*, 1983). The detailed arguments are too complex to be restated here, but in essence they suggest that (1) the polyoma experiments were seriously flawed, and (2) the implications about lack of hazard could not be generalised to other types of virus. The content of these papers was raised by Haber, one of the trade unionists on GMAG, at the final meeting of the Group, and it was recommended that the new ACGM should consider the matter further.

Irrespective of whether these criticisms are accepted, and of whether they imply any need to change the categorisation of experiments, Bartels points out, concerning the recent attitudes of scientists:

> the scientific community seems to have become rather wary of again involving itself in the debate on the hazards of gene manipulation. . . . One may compare the shrill voice of past

129

years levelled against charges of possible hazard with the deafening silence which prevails at present. (Bartels *et al.*, 1983:80)

In summary, the major change concerning possible risks of genetic manipulation has been in scientists' attitudes rather than in unequivocal new evidence. Professor Mark Richmond, a scientist-member of GMAG, commenting at the Wye conference on the proposal to relax the NIH guidelines – with which he agreed – stated:

[People ask] what has happened to make the Americans suddenly change the level under which experiments should be done from a very severe position to a much, much looser one? Has the actual amount of concrete evidence that has come forward been enough really to justify that? . . . I feel the main basis on which it has been done is only a change in the climate of opinion. I don't think you can really say that the amount of risk assessment work that's been done, and the amount of scientific realisation that has come to us, is solely responsible for the change in position. (Richmond, in Morgan and Whelan, 1979:229)

The legacy of the Williams Working Party

The Williams Working Party recommended that there should be four levels of precautions for the physical containment of any hazards arising from genetic manipulation research. Category I was the least stringent, and involved what is known as 'good microbiological practice', i.e. no mouth pipetting, smoking, eating or drinking in the laboratory, and so on. At the other end of the scale, category IV involved the kind of containment otherwise associated with the handling of highly infectious agents and required the use of, for example, totally sealed laboratories under reduced air pressure, air locks for entry and exit, and the sterilisation of all waste. The report also produced a table in which certain categories of experiment were matched with these containment levels (Williams, 1976:9–10). The text indicated that the classification of the possible risks was based on four factors. These were (adapted from Williams, 1976:6–7):

(1) The source of the DNA. DNA taken from organisms closely related to the organism at risk (e.g. humans) was taken to be more potentially hazardous than that from phylogenetically distant organisms (e.g. bacteria).
(2) The degree of purity of the DNA. Impure DNA would

contain unknown sequences, and these might prove hazardous.
(3) The host/vector system used. This related, *inter alia*, to the ability of the bacterium used to survive outside the experimental system.
(4) The experimental procedures used.

It was the first of these four factors which was to become the focus of problems in GMAG's categorisation activities, and this question will be dealt with in more detail shortly. For the present, it should be noted that any system of categorisation was beset by a number of difficulties. First, as already noted, the risks were all 'hypothetical', since there was no experimental data indicating any actual risks from recombinant DNA techniques *per se*. Second, the nature of the biological processes, and of knowledge of them, was, and still is, such that it was difficult or impossible for scientists to be able to state unequivocally that any of the putative hazard scenarios could not obtain.

In view of this fundamental uncertainty, the Williams report suggested that the classification table and the four factors should be taken by GMAG as the initial basis for its categorisation activities in order to allow it to 'get off the ground'. The intention of the Working Party was that this classification should be used flexibly as part of a 'case law' approach, accommodating to the expected rapid increase in biological knowledge of the supposed risks. So, allied to the more detailed suggestions, GMAG was enjoined to review the basis of the assessment of possible risks in the light of new knowledge, and as the situation demanded.[3] In effect, then, the Williams remit for GMAG comprised the details of a categorisation scheme, and a broader philosophy encouraging flexibility.

When GMAG began to receive and process applications, early in 1977, it followed the details of the categorisation table fairly closely. The Group soon settled down to the details of the Williams approach, and does not seem to have given much explicit attention to the basis of its categorisation activities. Of course, there were good organisational arguments for this. Scientists were waiting to proceed with experiments, and there were other pressing issues, such as the possible introduction of a medical monitoring scheme, the establishment of local safety committees in university laboratories, and the confidentiality issue, all of which demanded attention.

Whether these reasons were sufficient to justify the lack of consideration of the basis of categorisation is a matter of some debate amongst GMAG members. It has been argued, for example, that in the first year there was insufficient experience of the working of the scheme to warrant any change. A consequence, however, was that as the process of building up a body of case law proceeded, a number of difficulties with the Williams approach were encountered. There were anomalies or inconsistencies both between the categorisations of experiments which scientists outside GMAG saw as essentially equivalent in terms of possible risk, and between the British and American classifications of ostensibly similar experiments. The Williams approach came to be seen as lacking any scientific basis or justification. In addition, many scientists had become increasingly convinced that the initial fears about the hazards of genetic manipulation had been overstated, and that certain experiments were categorised at levels of containment substantially higher than was required. Thus, there built up considerable scientific pressure on GMAG against the absolute levels of containment for certain experiments, and against the anomalies of, and lack of scientific rationale for, the Williams approach.

These pressures were most strongly felt by the scientists on GMAG. The events which led to the change in the basis of categorisation occurred on a scientific/technical subcommittee of GMAG which was formally concerned with the validation of safe host/vector systems, the Safe Vectors Subcommittee. Brenner, who was responsible for the origins of GMAG's eventual response to those pressures, was a co-opted member of this committee.

The development of the new risk assessment scheme on the Safe Vectors Subcommittee

An immediately identifiable event in the evolution of this new risk assessment scheme was the December 1977 meeting of the eukaryote subgroup of the Safe Vectors Subcommittee. Part of the discussion at the meeting centred on the question of whether greater evolutionary distance of the source DNA from the organism at risk implied a greater or lesser degree of potential hazard. In the end, the only consensus seems to have been that, contrary to one of the major principles of the Williams report, there was no simple general rule here. At the following meeting of the subcommittee, in January 1978, Brenner made a verbal presentation of a multi-dimensional approach which took into account the 'consensus' of the previous meeting. In so far as the new approach developed from these previous discussions on the subcommittee, it is regarded by some as the origin of Brenner's scheme. Certainly, the scheme can be seen as a response to the previous, inconclusive, deliberations; but there was not, in fact, any simple cause and effect relationship. Brenner had, with others, been concerned for some time about the way in which GMAG had operated the Williams guidelines.

During the course of the controversy over the question of the regulation of genetic manipulation research, Brenner had been closely involved in a number of decisions about the way in which the possible risks or hazards were assessed. Earlier, in 1975, he was an influential member of the Asilomar conference; and also had some influence on the decisions taken at the important La Jolla meeting in America, at which one of the versions of the National Institutes of Health (NIH) draft guidelines were considered.[4] In terms of events in the United Kingdom, he had given evidence to the Ashby Working Party in 1975, and had been a member of the Williams Working Party in 1976. In the latter case, the four factors approach to the assessment of possible hazards, which formed part of the Williams remit to GMAG, owed a great deal to the work of Brenner and another British scientist, Professor Peter Walker of the University of Edinburgh. Indeed, Brenner recalls that the emphasis on flexibility, and, in particular, the manner in which GMAG was enjoined to reassess the basics of the approach to categorisation, were principles that he advocated strongly on the Williams committee. Thus, Brenner had for some time been involved in the consideration of the broad approaches to the assessment of the possible hazards of genetic manipulation, and had been influential in the formation of the Williams philosophy.

There is some additional background which should be noted here. Early in 1977, Brenner's laboratory at Cambridge had sent GMAG a 'proposition' detailing the manner in which they proposed to assess their experiments. The document, which was regarded by some GMAG members as being *ex cathedra* in tone, in effect carried the message that the details of the Williams approach should not be followed too closely. The particular rationale for what, superficially at least, looks like a modification of Brenner's stance on the Williams committee is not of direct concern here. The mere fact of a difference between GMAG's and Brenner's readings of Williams is, however, relevant to later events.[5] One final point on Brenner's involvement is that he was also chairman of a short-lived organisation known as the Genetic Manipulation Users Group (GMUG). This was set up under the auspices of the Medical Research Council to provide a forum in which the interests of the users, scientists performing or likely to perform such experiments, could be represented. During the two meetings of this pressure group in the summer of 1977, a series of criticisms of GMAG's operations were made, many of which bore on the appropriateness of the Williams approach to categorisation.

In short, Brenner, like many scientists, was in a position to appreciate the growing anomalies and contradictions that scientists saw in GMAG's operation of the Williams approach. However, unlike many scientists, he was in a position to do something about it. Although there had already been some extension of the Williams approach, Brenner felt that there could be no real progress without rethinking the basis of categorisation from the beginning. The extent to which the scheme that Brenner produced should be regarded as a new or radical departure from previous assessments, especially that of the Williams report, is a matter of some debate, and will be returned to later. It is worth noting at the outset, however, that in his initial presentation to the Safe Vectors Subcommittee, Brenner emphasised that though the details of his approach to the analysis of risks, and hence the categorisation of experiments, would differ from those of the Williams report, it should be seen as a development, and as compatible with the broader Williams philosophy which enjoined GMAG to reconsider the basis of assessment as the situation demanded.

Brenner agreed to produce a written paper for the Safe Vectors Subcommittee.[6] This came in three parts, the first two in February, and the third in May 1978.[7] Before examining the basis of the approach detailed in the paper, it will be useful to look at an example of the kind of difficulty associated with one of the four factors in the Williams approach. In the Williams scheme, and the principle is also to be found in the Asilomar approach (see Krimsky, 1982:185), an important factor in the assessment of the potential or possible risk was that of phylogenetic relatedness.[8] To summarise, the possible hazard associated with a sequence of DNA was taken to be inversely related to its distance on the evolutionary scale from the organism at risk. Other important factors would, of course, need to be taken into account, but *ceteris paribus*, bacterial DNA was taken to be less potentially hazardous than that of, say, reptiles, which, in turn, was less hazardous than mammalian DNA. The difficulty can be illustrated if we take the case of the segment of DNA which codes for histone, a type of protein. The DNA, and the histone, are almost identical across a very wide range of species, including humans. One scientist explained during interview:

'A histone gene in you looks much the same as a histone gene in a pea. If we took your histone gene and put it in a plasmid it was category III. If we took a plant histone gene and put it into a plasmid it was category I. Now that was frankly ridiculous, since the genes were the same. And once it was DNA it was exactly the same, and no-one could tell whether it came from you or a pea. And therefore this intellectual dilemma was very, very, acutely felt . . . by everyone on GMAG.'

By comparison, the Brenner scheme approached the assessment of the possible risk of an experiment through the consideration of the potential outcomes of particular cases of genetic manipulation experiments. In outline, the new scheme revolved around a series of canonical *Gedankenexperimenten* which would, on the basis of existing scientific knowledge, lead to a particular, known, hazard or outcome. Actual experiments, and hence actual outcomes or hazards, could then be assessed by comparison with the canonical benchmark. In assessing any proposed actual experiment by comparison with the canonical reference standard, the scheme would be concerned with the specific biological activities of the DNA involved, and a consideration of the particular conditions under which these activities might, or might not, affect the organism at risk. In terms of the histone example, the Brenner scheme would assess the possible hazard on the basis of the known biological activity of the DNA, not on whether it came from a human being or a pea. Since the DNA in question, and hence the biological activity, was the same irrespective of the source, the categorisation would be the same in both cases.[9] It should be noted that the present point is concerned, not with the absolute level of classification, but with the manner in which the Brenner scheme offered a way out of the inconsistencies of the Williams approach.

The tripartite paper which Brenner presented to the Safe Vectors Subcommittee was pedagogic in style. It laid out a set of assumptions, defined the factors to be taken into account in the assessment of outcomes, and developed the logic of the approach. One 'assumption', or 'proposal' in Brenner's terms, clearly stated at the outset, was that consideration of the effects of a genetic manipulation experiment on an organism should be restricted

to those outcomes for which natural paradigms of pathogenesis already exist. . . . We are asserting that if recombinant DNA molecules have adverse effects, these will be subject to the same biological constraints as naturally occurring ones, and will be very much like them.[10]

The clear implication is that any possible risks associated with genetic manipulation experiments are not different in kind from those of other forms of biological experimentation. In effect, this assertion represented an answer to one of the major questions and concerns which led to the Asilomar conference: 'would the new genetic manipulation techniques lead to new or different kinds of hazards?' Some scientists had always felt that the answer to this question was in the negative, but as already noted, the structure of molecular biological knowledge is such that it did not, and still does not, permit an unequivocal 'No'.

The fact that Brenner could 'assume' or 'propose' such a statement is an index, and a consequence, of increased scientific familiarity with the area, and of changes in scientific opinions about the likelihood of risk. The view which the assertion expressed was not original to Brenner for it was, by then, becoming increasingly apparent to many scientists. Nevertheless, by presenting the view in a rather bald fashion, and without the qualifications and caveats sometimes found, Brenner brought it 'out of the closet'. It should also be noted here that some such 'assumption' or 'proposition' is required for the analysis of experiments in the Brenner scheme. The comparison of actual experiments with canonical or standard experiments was to be carried out on the basis of existing scientific knowledge, and this could only be done if, in principle at least, the biological processes in genetic manipulation experiments were the same as those known to operate elsewhere.

A further point of interest about the content and presentation of the Brenner scheme is that it changed what might be called the 'rhetorical ground' on which the assessment of risk takes place. As one aspect of this, Brenner ruled out of court hypothetical outcomes which were not constrained by existing biological processes. A related aspect can be illustrated through the propensity of some of those involved in the early debate on genetic manipulation to construct scenarios of possible hazards. An example might be, 'Suppose *E. coli* acquired the ability to produce a toxin harmful to humans?' Large numbers of such possibly hazardous scenarios can be constructed with ease, especially so if the biological possibility of such eventualities occurring are unquantified, or assumed to be unquantified.

Seen through the perspective of the Brenner scheme, the Williams and Asilomar approaches are unsatisfactory in that they attempted to apply a set of containment procedures to a subset of these scenarios without first addressing the question of the biological possibility of the outcome postulated. The outcomes are in some sense feasible, but the assessment did not proceed by way of a detailed analysis of biological processes, but rather through the simplification of the four factors thought to relate to the severity of the outcome. Williams suggested how a possible untoward outcome might be avoided by the adoption of certain precautions. That is, through the use of disabled strains of *E. coli*, by the use of physical measures to contain any toxin-producing bacteria, and so on. But the likelihood of the outcome was not, for Brenner, assessed with any degree of precision. In the Williams approach there is a tendency to take the outcome as granted, and then apply the appropriate precautions. The rhetorical implication of this is that any imperfection in the containment procedures will allow an escape, which would lead to the untoward outcome.

In a sense, the Brenner scheme looked at particular scenarios and asked what conditions would have to be fulfilled for an outcome actually to occur. For example, Brenner analysed all the biological aspects of an experiment specifically designed to create a toxin-producing bacterium, and to get it to survive in the human gut. The analysis is then potentially concerned not only with the molecular biological aspects of the gene transfer, but also with questions of the nutritional requirements of the organism, the pH of the gut, competition from other bacteria, and so on. Aside from the obvious scientific value of this approach, it also carries a certain rhetorical force. The emphasis is on all the various biological steps and transfers that have to occur, in the correct sequence, in order for this experiment to work and produce a hazardous outcome.[11] Under Williams, all the various steps had to occur for a non-hazardous outcome. The difference can be thought of as analogous to, but not the same as, the distinction between fail-safe and fail-dangerous systems. Brenner's scheme does not, of course, make any difference to the performance of any biological system; rather, it forces attention on to the biological fail-safe characteristics of the experimental system being assessed, in addition to the fail-dangerous aspects of containment procedures. In addition to these rhetorical or persuasive factors an important innovation of the Brenner scheme was the attempt to specify estimates of the probability of a given outcome occurring. Each outcome is regarded as the final point of a generation pathway which consists of the various necessary biological steps. If a probability value could be assigned to each of these steps, then the overall probability of risk could be derived. If it is assumed for the moment that the relevant biological data is available, then a probability value can be established for any one of the canonical or reference standard experiments. Actual experiments which differ from the reference standard in known ways can then be compared in an orderly and consistent manner. Indeed, if this could be done, it would lead to quantitative statistical analysis of the kind found, for example, in the assessment of radiation hazards.[12]

In summary, the Brenner scheme held a number of promises and attractions for the Safe Vectors Subcommittee. It obviated the troublesome Williams principle of phylogenetic relatedness and its consequent anomalies; it proposed an ordered consideration of generation pathways based on existing biological processes; and it held out hope of some form of more conventional, quantitative, risk analysis.

As noted above, Brenner's developing ideas for a new approach to the assessment of experiments were first mentioned briefly at the Safe Vectors Subcommittee meeting in December 1977, and he then made a verbal presentation in January 1978. At this early stage Ravetz, a public interest representative with a long history of concern about the social implications and responsibility of science, was already enthusiastic about the possibilities for the further development of the verbal outline.[13] After the verbal presentation, and during the period in which the three parts of the paper were produced in February to May 1978, Brenner became the focus of a small group of Safe Vectors Subcommittee members who were increasingly confident about the potential of the new approach. Indeed, Ravetz formally expressed his view that the scheme was worthy of further development directly to the chairman of GMAG, Wolstenholme, as early as January 1978. Given the remit of the Safe Vectors Subcommittee (the validation of safe vector systems), the agreement of the chairman, Subak-Sharpe, for this digression was important, as was that of Wolstenholme.

During the period from May to June 1978, the emergent scheme was not debated formally on GMAG but remained notionally within the Safe Vectors Subcommittee. However, interest was not confined to the appointed members of the Safe Vectors Subcommittee, and its meetings were in any case open to any member of GMAG. Maddox, a public interest representative, and an ex-editor of *Nature*, and Williamson, a professor of biochemistry at St Mary's Hospital Medical School, London and a trade union nominee, have both been identified by some participants as supporters of the principle of the scheme.

In June 1978 two meetings were arranged so that the Safe Vectors Subcommittee could consider its response to Brenner's tripartite paper. By this time, the detailed arguments of the scheme were fairly well known on the Safe Vectors Subcommittee and the approach broadly accepted within the group comprising the Safe Vectors Subcommittee members and the other interested GMAG members noted above. It seems that the small network centred on Brenner had been involved in correspondence and other contacts. This led to Ravetz producing a short non-technical paper which introduced the scheme, and to the production of an important annex by Rigby, a research scientist and co-opted member of the Safe Vectors Subcommittee. Rigby's contribution was a series of worked examples of the generation approach, based on actual experiments, which attached probability estimates to the various biological steps. The great value of this was the demonstration that existing knowledge was, in some cases at least, sufficient for the derivation of viable estimates of overall risk.[14]

The particular event most frequently cited in participants' recollections of this period was a meeting of the Safe Vectors Subcommittee held in Glasgow. It was here that Rigby presented his examples, and thereby helped to show that the scheme would work. It seems that Subak-Sharpe, the chairman of the committee, had hoped that the scheme would be tested by asking its members to work through a series of examples independently. Obviously, if everyone obtained the same result this would be an important validation. As it happened, no formal test took place, although in a significant sense the Glasgow meeting was partly a 'teach-in' on how to use the scheme, since the assembled scientists initially produced comparable categorisations, but on the basis of quite different probability estimates for the various biological steps. There was now pressure to proceed as rapidly as possible.[15] The Safe Vectors Subcommittee proposed the Brenner scheme to GMAG, and the Group considered the proposal at its July 1978 meeting. The paper tabled, in essence the Brenner/Rigby paper, was accompanied by notes from Ravetz and Maddox. Maddox made a number of detailed critical comments, and also emphasised the importance of the manner of the presentation of the scheme to the scientific and general public. Maddox pointed out the need to take account of the social and political implications of proposing a major change in the basis of the categorisation of genetic manipulation experiments. The package of material for GMAG's consideration was completed by a short covering paper. This rehearsed the problems of applying the Williams approach, and reiterated the theoretical innovations of the Brenner/Rigby paper, whilst also emphasising the continuities with the Williams philosophy of reassessment. In addition, the covering paper outlined a possible timetable of events for the following six months which GMAG 'might like to consider'. It was suggested that the various papers could be 'consolidated and amplified', and that the revised version could then be considered by GMAG prior to its presentation to the scientific community. Following consultations with the appropriate official bodies, the new basis for categorisation could then be introduced, and a meeting held to 'educate' scientists in the use of the new scheme.

Broadly, GMAG welcomed the scheme, despite a number of reservations, and some not unnatural difficulties in grasping the full implications of the new approach. In particular, GMAG felt that further development of the scheme was required, partly to take into account the kind of criticisms voiced by Maddox, and also in relation to the way in which the probability values were assigned to the nodes or the generation pathways. It was decided that a small working party should be set up to produce the revised version. This was known as the 'Working Party to Consider the New Guideline Criteria'. Wolstenholme then appointed Maddox as chairman, with three scientists as members, Richmond and Dumbell who were GMAG members, and Weiss, a co-opted member of the Safe Vectors Subcommittee.

A sense of urgency had been injected into GMAG's consideration of the basis of its categorisation activities, which contrasts with the somewhat leisurely approach which had been adopted during the first half of 1978. The reasons for this are not hard to find. Within the British scientific community affected by GMAG's regulations there was a growing frustration with some of the anomalies that the Williams approach had produced, but more importantly, perhaps, with the absolute level of containment attached to certain types of experiment. Also, during the middle part of the year it became clear that the American Recombinant DNA Advisory Committee of the NIH was proposing a significant relaxation of its containment categories. This would probably affect most European countries as well, for they had decided to follow the NIH approach, rather than the British model. Potentially, Britain could be out on a limb, and researchers might well go abroad, which in turn would ultimately affect not only the standing of British science, but also the likely benefits of the expected emergence of a major biotechnology industry using genetic manipulation techniques.

There was, then, a nexus of pressures which can all be seen to encourage urgent action. One further factor also deserves comment, and this is the considerable pride members took in what they saw as the achievements and successes of GMAG. It was often stressed, somewhat chauvinistically perhaps, that in distinction to the American approach, Britain had gone its own way and produced a more coherent, and more strictly implemented, system. That system was now in danger from the anomalies it had produced, and the Brenner scheme offered a potential way out of the difficulties. Undoubtedly, the likely American moves were influential in dictating the time scale of GMAG's actions, and it was also important that these actions should not be seen merely as a response to an American lead.

The Working Party to consider new guideline criteria

The newly established Working Party on the guideline criteria held a meeting in July, when they decided to look more closely at the numerical and statistical aspects of Brenner's scheme. In addition, the chairman, Maddox, produced an initial paper which expanded on and questioned some aspects and assumptions. This was circulated to a number of interested parties, including Brenner, for comment. There followed a period of intense activity in which Maddox especially was involved in correspondence and meetings with Brenner and others. Apparently, there was much informal committeeing, lunching and exchanging of drafts. The outcome of these negotiations was a paper which was laid before the October 1978 meeting of GMAG. Following its comments a final version was prepared for the November meeting, and was then approved.[16]

Subsequent actions of GMAG continued roughly in accordance with the timetable which had been suggested at the July meeting. A version of what was now called the 'Prospective new basis of categorisation' was circulated to official bodies, and published under a different title in *Nature* (GMAG, 1978). As it happened, few comments were received before the new scheme was presented to a meeting of scientists and members of local genetic manipulation safety committees in December. In 1979, GMAG began to introduce the scheme with a newly established Technical Panel whose responsibility was to oversee the operation, and to advise scientists on the mechanics of producing categorisations from it. Initially, the new mode of categorisation was operated in parallel with the Williams approach. The rationale for this tandem operation will shortly become clear in comparing the implications of the Brenner/Rigby scheme with the Maddox approach to categorisation which GMAG finally adopted.

Brenner's section of the document placed before GMAG in July

1978 has been described by its author as an essay in theoretical pathogenesis, and as a piece of risk assessment. The discussion which follows will first be concerned with what might be called its scientific aspects. It was noted earlier that the paper was pedagogic in style, setting out a series of assumptions and following through the logic of an analysis of the possible risks of genetic manipulation experiments. The approach is based on the identification of those generation pathways, or generation trees, which could, biologically, lead to specified outcomes. In part, the argumentative structure seems designed to force the reader away from the four factors approach of the Williams report, and towards a consideration of the biological mechanisms and processes necessary for the creation of a pathogen or a hazard. That is, the reader's attention is directed towards areas of agreed biological knowledge, and away from speculative scenarios.

Examination of the content of the paper indicates that little new data or information is presented. Rather, the importance of the paper is to be found in the reconceptualisation of available data. Brenner draws upon a variety of existing cultural resources and brings them together in a new configuration. This is not meant to imply that the scheme is derivative in any pejorative sense of the word, but rather that it has certain characteristics typical of scientific innovation. One of the scientists on GMAG, frustrated by the illogicalities of the Williams categorisations and by the complaints of his colleagues to the point of considering resignation, commented during interview that Brenner's analysis

'was the most important watershed. . . . I can be critical of Sydney Brenner, but the truth is that his analysis made it clear that it was possible to delineate a set of experiments that posed no risk . . . that's meaningful. And the ability to categorise experiments in a meaningful way . . . as opposed to a, how can one put it, didactic way according to a schedule, was extremely appealing to everyone on GMAG, whether you talk about public representatives . . . or anyone else. The reason it was so appealing was that it made sense. All of a sudden you were really looking at risk parameters . . . you were not looking at species [the phylogenetic relatedness argument] any more . . . not arbitrary headings, you are looking at genuine risk parameters. That was really a change in our whole approach.'

In short, the Brenner scheme can be regarded partly as a piece of science. It draws upon scientific and cultural resources to produce what is reasonably identifiable as a scientific rationale for a new approach to the analysis of the possible risks of genetic manipulation experiments. But, of course, the scheme was also developed to get GMAG out of a predicament, by providing a coherent basis for the administrative and practical tasks of allocating experiments to particular categories of containment. In other words, it was also intended as a piece of science policy. The most useful way of bringing out this aspect is to look at the ways in which the Maddox Working Party responded to the implications of the Brenner scheme, and how they changed it.

Brenner's paper had introduced the idea of generation pathways. On the basis of existing biological knowledge, probability factors could, in theory, be assigned to the various transition points or nodes in such pathways. This would then lead naturally to the ideal of quantitative risk analysis. The Maddox Working Party, however, doubted that existing biological knowledge was sufficient to bear the burden. Indeed, they suggested that it 'will be some time, perhaps several years, before biological understanding and quantitative data have accumulated sufficiently for such techniques to be applied to all proposals for experiments in genetic manipulation' (GMAG, 1978:105),[17] and, 'For the time being . . . such a procedure would provide a spurious sense of precision' (GMAG, 1978:107). They also pointed to a further serious obstacle to any form of quantitative risk assessment in the area. Under the Williams approach, and GMAG, there were four levels or categories of containment, CI to CIV. As the Working Party put it: 'it is proper to associate with each category of containment a specific probability . . . that organisms will escape' (GMAG, 1978:105). Obviously, this is a vital aspect of any assessment of the overall risk of an outcome. Unfortunately,

> there are at present only rough estimates of the ratio CIV/CI, the relative safety of the most stringent and the least stringent of the categories now in use. The ratio is likely to be less than 10^{-6} (corresponding to a factor of 100 for the interval between each category of containment) and may be less than 10^{-9}. (GMAG, 1978:105)

Brenner's analysis had adopted an estimate of about 10^3, or a factor of 1000, between usefully identifiable levels of possible risk. The Working Party concluded that it was 'important that the true range of safety offered by the four categories of containment should urgently be defined more precisely' (GMAG, 1978:105). In this way the Working Party came out against the possibility of any immediate, general introduction of quantitative risk analysis. But they did not abandon the Brenner approach.

Reading the papers of the Maddox Working Party, and in particular the version which GMAG finally approved and published in *Nature*, it is clear that, in principle, they accepted the generation pathway approach that Brenner had outlined, even though what many considered to be Brenner's more scientifically persuasive paper was not reproduced in the *Nature* version. GMAG's publication emphasised the value of Brenner's scheme in making possible an objective and consistent analysis, and it strongly underlined the quantitative analysis of risks as an ideal. In a formal sense the caveats expressed concern the limitations which available biological knowledge placed on the implementation of a programme of quantitative analysis. According to taste, it looks as though GMAG is more hesitant, or Brenner more optimistic, about the ability of scientists to provide experimental evidence on which to base quantitative analysis. On the other hand, there seems to have been a body of opinion on GMAG to the effect that more rough-and-ready sets of numbers or probability factors could be applied which would lead to categorisations that were, in fact, reasonable. That is, there seems to have been a kind of tacit knowledge-based assessment of experiments which was felt to be 'right' in some sense, but which could not properly be shown to be justified in terms of the accepted rules of science. From some scientists' point of view, then, there was a problem of presentation; they were not able, within the Brenner scheme, to lay out the evidence for what seemed to them reasonable categorisations.

It seems useful to regard the Brenner scheme as a programmatic statement outlining an approach to risk assessment. It is exemplary, and perhaps, loosely, paradigmatic, in the Kuhnian sense of these terms. On this view, the reservations of the Working Party serve as reminders that the solutions to the technical puzzles of the research programme, that is, particular risk assessments, would require scientific time and effort.[18] GMAG, however, had little time to play with. Widespread disaffection with the phylogenetic basis of its system of categorisation, pressures for relaxation of the guidelines following the USA lead, and the administrative task of having to continually process an ever-increasing number of research applications as the field grew meant that GMAG had to act without delay. The Working Party, therefore, made a series of proposals which were accepted by the Group.

GMAG's approach can be summarised in the following manner. The containment categories had been physically and organisationally institutionalised, and were thus difficult to change. As part of the new approach, four 'starred' categories to be known as CI*,

CII*, CIII* and CIV* were introduced which were identical to the Williams categories in terms of physical containment. The old Williams categorisation scheme and the new risk assessment scheme would operate together, with experiments being allocated under one or the other scheme. Those experiments for which some kind of risk assessment was possible would be allocated to the starred categories, otherwise categorisation would continue as before.

In allocating research proposals to the starred categories, the *Nature* (GMAG, 1978:106) paper stated that only 'qualitative' or 'order of magnitude' estimates of the transition probabilities would be used in most cases. GMAG would consider three important factors in the generation pathway, the Access, Expression and Damage factors, when determining the category to which a given experiment would be allotted. The Access (A) factor referred to the overall probability that a DNA fragment or gene inserted into a bacterium, for instance, could be transferred to the organism at risk. Expression (E) was an estimate of the probability of a segment of DNA actually being expressed by the cell to produce a harmful product, for example, a toxin. The Damage (D) factor estimated the probability of an expressed gene finally having a harmful effect. However, categorisation decisions would, for the most part, 'continue to depend on judgement rather than calculation' (GMAG, 1978:108) though researchers could, if they so wished, submit with their proposals evidence or data which might bear on a numerical determination of the A, E and D factors, and hence on the risk category. In order to give this new basis for GMAG's categorisation procedures some structure, the *Nature* paper also proposed that:

> experiments with which researchers are familiar should be allocated to the four containment categories [the starred ones] on the basis of a *qualitative* and *relative* assessment of their risks; and that the potential risks of novel experiments should then be assessed in relation to these familiar paradigm experiments. (GMAG, 1978:107, emphasis added)

Clearly, the compromise of using the Williams phylogenetically based system of categorisation in conjunction with the risk-prob-ability-based 'starred' categories when data was available differed markedly from the Brenner scheme as it was originally proposed to GMAG. The Brenner scheme had promised sound scientifically based risk assessment procedures which were canvassed as 'objective' and 'consistent', which were based on 'existing biological knowledge', and which would allow 'quantitative risk analysis'.

As already suggested, in important aspects the Brenner scheme was programmatic, and was implicitly treated as such by the Maddox Working Party. The ideal of the Brenner system, or, at least, the rhetoric that had become associated with it, was carried over into the *Nature* paper and presented as characteristic of GMAG's proposals for its longer-term categorisation activities. The shorter-term tandem operation of the two systems of categorisation was then proposed as an approximation to this ideal. This allowed for the progressive transfer of the familiar experiments to the starred categories where they could act as paradigms for novel experiments, with the risk assessment being based on judgment, rather than on quantitative procedures.

In effect, this form of presentation allowed GMAG to borrow legitimacy from the programmatic ideal, whilst facing up to the current difficulty of actually implementing the programme. The operation of the two systems also had the advantage that it allowed GMAG to remove known inconsistencies in categorisation, such as the anomalies caused by the Williams phylogenetically based approach. Flexible use of the two systems would mean that consistency of categories could be maintained, albeit at the expense of some limited '*ad hoc*ery' in the choice of the appropriate categorisation system.

The new risk assessment scheme was published in *Nature* in November 1978 and, as part of the relatively short period of consultation, was presented to a meeting of the members of the local biological safety committees just before Christmas. The published version maintained the 'party line' that the new system represented an evolution of the Williams principles, and did not explicitly mention the topic of lowering the containment levels for experiments. The meeting of scientists was characterised by expressions of doubt about the possibility of providing sufficient scientific knowledge to make the scheme work and there were also more specific complaints about the categorisation of self-cloning experiments, as noted in chapter 4.

In February 1979 an *ad hoc* group of the Royal Society published a strong critique of the risk assessment scheme:

> We believe that the discussion of such conjectured risks should be adequately documented by reference to the scientific literature and subject to comparable standards of scientific scrutiny.
>
> The discussion in the GMAG paper falls far short of these standards. (Royal Society, 1979:509)

Commenting on the 'open, documented and subsequently

published consultation with many groups in the United States and abroad', the Royal Society urged that GMAG follow the American example, and that they should abandon the case law approach in favour of 'set categories for given types of experiments' (Royal Society, 1979:510).

On a strict definition of 'scientific', the criticisms of the Royal Society had some force. In effect, the new risk assessment scheme left the containment categories as they were, and proposed a rationalisation of the process of allocating experiments to these containment levels. In some cases a reasonable scientific argument could be made for the (approximate) figures used, but in other cases, and especially concerning the 'Damage' factor, the scheme was forced to rely on 'average' values of damage for known pathogens, and there was no real certainty that these would in fact apply. A more telling criticism, and one not mentioned by the Royal Society, was that whatever the admitted formal deficiencies in the process of arriving at categorisations for experiments, the different degrees of protection of the four levels of containment into which they were allocated were – and still are – largely unknown.

GMAG noted the Royal Society critique, but did not publish a response. A defence given in interviews was that the criticism missed the point. Part of the purpose of the new risk assessment scheme was to find a more or less consistent means of categorising experiments as a prologue to effective relaxation. Also, GMAG was in the business of advising scientists on containment levels. It had to have defensible reasons for those levels, defensible reasons for changing them, and it simply was not possible for the Group to announce that certain experiments were uncategorisable.[19]

GMAG went ahead with the tandem operation of the Williams and risk assessment approaches, and later dropped the former in favour of the latter, as described in chapter 4. Despite trade union concern over the possible surrender of the power of the Group to a subcommittee, the Technical Panel engaged in informal negotiations with proposing scientists, and introduced a gradual easing of restrictions. At certain points this made conditions somewhat less stringent than the NIH guidelines. The major change in the basis of categorisation had been successfully negotiated, and GMAG did not again have to face such hostility from scientists.

Discussion

Particular aspects of the evolution and development of GMAG's response to the problems of categorising experiments on the Williams system have been discussed in some detail. They can be described as a series of mainly scientific, but partly administrative, decisions in an institution specifically charged with the making and implementing of an area of science policy.

During the period concerned, GMAG was faced with the problem of achieving an administratively workable system of categorisation from one whose basis was under serious scientific criticism. Moreover, it was during this period that pressures for relaxation of controls on genetic manipulation experiments were mounting in all countries where research was taking place. The level of both scientific and public apprehension concerning the likelihood of hazards was reducing. The benefits of research were coming to be more widely seen as outweighing the risks. The potential technological applications of research in the medical, agricultural and industrial fields were being recognised in plans for scaling-up experiments in pilot plants, and then production, in the development of biotechnology. From the point of view of the United Kingdom scientists, pharmaceutical and agricultural companies engaged in genetic manipulation research programmes and research councils financing university-based research, relaxation of controls, at least in parallel with their relaxation in other countries, was seen as crucial to maintaining international competitiveness.

The initial proposal by Brenner of a risk assessment scheme whose eventual effect would be to place most proposed experiments in lower categories with less stringent, less costly, and more widely available containment requirements, and its adoption in hybrid form by GMAG, provided a mechanism for facilitating relaxation of the regulations. GMAG maintained a remarkably high degree of unanimity about the benefits of the new approach. The climate was conducive to the reception of a scheme bringing about relaxation, but it was also demanding of one which provided an acceptable, sound scientific basis to any new system. Undoubtedly, had the Brenner scheme not been proposed, then relaxation of the regulations would have taken place, possibly as happened in other countries, on what has been seen, with both supportable and jingoistic satisfaction by most GMAG members and scientists in the United Kingdom, as a less rational basis.

Analytically, there are two important aspects to the introduction of the new approach to the categorisation of experiments. The first aspect involved a reconceptualisation of the possible basis of GMAG's categorisation activities during the discussion on the Safe Vectors Subcommittee. Major factors here were that any approach should be scientifically defensible, and should lead to defensible categorisations. Undoubtedly, it was equally desirable that some effective lowering of containment levels should eventually emerge, even though these could be presented as not following necessarily from the new scheme.

Commenting on this first aspect of the new approach to categorisation, a number of those interviewed drew a distinction between the 'scientific' and the 'social' or 'political' dimensions. The scientific dimension of the scheme concerned the logic, and the consistency in producing relative levels of risk and appropriate containment levels; it was then a matter of 'social' or 'political' decision as to what the level of containment should be in absolute terms. (This kind of distinction appears to be fairly frequently found in risk analysis research according to Johnston, 1980.) It is important to note that the scientific/social distinction was not presented in terms of 'true' versus 'false' knowledge or of 'legitimate' versus 'illegitimate' factors, but as equally legitimate and necessary dimensions of the new policy. While the borderline between the scientific and the social was not the same for all those interviewed, the recognition of the equality of the two dimensions was.

The second aspect relates to much of the work of the Maddox working group, and also to the later stages of the considerations of the Safe Vectors Subcommittee. As Brenner's theoretical scheme came under closer scrutiny and was fleshed out with examples of quantitative estimates, the difficulties of providing scientifically based quantities defensible in terms of current scientific knowledge became apparent. As Johnston (1980) has pointed out, logic tree or logical path analysis in risk analysis research quickly becomes logistically impossible due to the potential number of pathways and simplifying assumptions which have to be made. That is, by the idealised standards of physics on which risk analysis research is based, the latter is forced to compromise in the face of increasing complexity. But even in terms of the existing practices of the 'immature' field of risk analysis research, current knowledge of the biological processes involved in genetic manipulation was insufficient as a basis for quantitative, probability-based estimates of risk. Hence, the Maddox working group was led to move away from the Brenner ideal of a quantitative scheme towards 'order-of-magnitude' estimates, based on scientists' tacit knowledge.[20]

Thus the second aspect of the development of the new categorisation policy reveals a shift away from the ideal of a mechanical system which would automatically lead to a relative scale of risks. Paradoxically, perhaps, this does not mean that there was a greater degree of input to the decisions about categorisations by 'social' factors, but rather that scientists' potential control was actually increased. The reason for this is that, to the extent that such a mechanistic scheme can be produced, there are no scientific decisions to be taken, since the eventual level of risk follows automatically from the data. On the other hand, to the extent that numerical data are not available, decisions have to be made by scientists about the approximate orders of magnitude which are to be used in the analysis. Thus the lack of agreed or consensual knowledge about the relevant parameters for risk analysis in practice led to a closer control over the risk categories, as illustrated by the negotiations on the Technical Panel.

In a general sense, the details presented here are consistent with the readings of those who have seen in these issues an overall tendency whereby social, political or ethical issues have become transformed into technical matters. However, the discussion also shows some of the exceptions to the overall picture, and, more importantly, indicates in detail points of entry, and possible entry, for such non-scientific factors. In this context it is tempting to consider an explanation in terms of the 'interests' approach, either of the 'Who benefits?' type, or the more explicitly developed notions of Barnes (1977). But, as Yoxen has put it,

> the structure of interests [on GMAG] is rather complex, with several of the academic biologists having financial interests in the expansion of the field, and two of the trades unionists being concerned with recombinant genetics as a research topic. (Yoxen, 1979a:231)

Indeed, this analysis suggests that the situation is considerably more complex. At first blush, it might be expected that the disciplinary interests of the scientists most closely associated with research would lead them to be supporters of the Brenner scheme, especially to the extent that it implied a relaxation of containment levels. But, in practice, this is not visible and, indeed, the public interest representatives Ravetz and Maddox are found to be clearly identified as early sympathisers with and protagonists for the general ideals of the Brenner scheme. Later, Maddox was influential in articulating the limitations of the scheme. Again, Subak-Sharpe, chairman of the Safe Vectors Subcommittee and a scientist closely connected with the research, was sometimes credited with having delayed the presentation of Brenner's tripartite paper by insisting on thorough consideration and checking. Subak-Sharpe's reasons included the importance of maintaining his scientific reputation for carefulness, and the validity of the proposed scheme. Brenner felt that the scheme would have gone through earlier if the public interest representatives had had greater influence on GMAG, and scientists rather less. In general, then, an examination of interests tends to break down before a detailed analysis because there are too many anomalies, and can only be sustained in its much more general, and perhaps less useful, form which sees the main 'interest' of GMAG members as easing the paths of both scientists and industrialists in the application and development of a highly fruitful, not to say potentially profitable, new technique (Yoxen, 1983).

More usefully, it is interesting to outline the differences between the British and American responses to the change in status of hazards. In the United States competing experts argued their cases in an open arena, while in Britain a committee of experts 'reviewed' the arguments. In both cases, however, the revised categories depended ultimately on scientists' tacit knowledge-based assessments of relative possible risks.

The nature and membership of the Recombinant DNA Advisory Committee was different from GMAG (see Krimsky, 1982: Chapter 11, *passim*). Like many American policy making committees, it was open to the public, published its minutes, and allowed a wide range of opinion to be heard. GMAG members, on the other hand, met in closed committee, were signatories to the Official Secrets Act, and produced occasional reports rather than published minutes of meetings. Clearly, such differences affected the ways in which evidence and arguments concerning potential hazards were considered.

In addition, while the public interest and the trade unions were represented on GMAG, and the latter were not represented on the RAC, the broader health and safety concerns were aired at Congressional hearings, and dealt with by local initiatives such as the 'citizens' court' and by local biosafety committees, as discussed in chapter 6 (see Krimsky, 1982:294 *et seq.*). Together with the more open approach of the Recombinant DNA Advisory Committee, these tended to demystify the nature of any differences concerning the interpretation of risks and hazards given by scientists. This is in sharp contrast with British procedures which tended to obscure the decision making process leading up to the acceptance of the new basis for categorising risks.

This chapter has shown how these changes occurred, and, in

doing so, has pointed to the close relationship between social and scientific factors in the policy making process. The British approach formally promulgated the traditional view that scientists operate mainly in the factual arena, and the nature of the institutional environment within which decisions were made did much to maintain it. However, the process of reconceptualising risks and hazards in recombinant DNA research was clearly a much more complex and negotiated process, relying on power, influence and rhetoric as well as science. As has been stressed, this is not meant to imply any untoward motivations on the part of scientists, or that matters were not fully and openly discussed. Counter-intuitively, as the new policy moved progressively away from the scientific ideal with which it began, it laid an increasing emphasis on the importance of the possession of scientific knowledge for its operation.

6
Science and public participation

During its early development, the potential benefits and possible hazards of the genetic manipulation techniques were recognised by a number of scientists. Some of them publicised their apprehensions, so initiating the often turbulent debate. Those eminent scientists involved in the call for a voluntary moratorium on certain types of experiment saw it as a temporary measure, pending the solution of essentially technical questions about the containment of possible biological hazards. However, the fact of such an unprecedented action, and the articulation of possible scenarios of hazard by other, eminent, biological scientists provided a condensation point around which generalised social fears about the implications of science and technology could crystallise into a 'moral panic'.

Thus wider social, ethical and political questions became a part of what had been originally intended as a technical debate about safety. In the subsequent controversy, issues relating to the social control of science and technology were indeed given an airing. But, as has been described, and perhaps inevitably, the very establishment of regulatory machinery to deal with the genetic manipulation 'problem' was itself based on the perspective that the main areas of decision making would be technical in character. Public concern was, however, sufficient to ensure that some mechanism for public accountability, that is to say some degree of social influence if not of social control, was embodied in the regulatory machinery. This chapter first outlines some of the general factors in the debate about the relationship between science and society which form the backdrop to the controversy. It then focuses on the issue of providing an input from the non-specialist into the arena of technical decision making, and in particular the British experience of representing the 'public interest'.

The context of public participation

The members of the Berg committee felt that there was sufficient scientific uncertainty about the possibility of dangers which might result from some uses of the new genetic manipulation techniques to warrant a pause during which the likelihood of hazard could be assessed. The immediate context of the call for a moratorium was an act of social responsibility in the face of scientific uncertainty. Any complete analysis of the broader context would need to consider a wide range of inter-related factors both inside, and external to, molecular biology, and which are related to important general perspectives on the role of science in society. In outline, there are ideological factors involved in the views of many scientists, and much of the wider public, that science is, ideally, an objective search for knowledge which should be encouraged to flourish unhindered in a free society. There are those social factors related to the changing role of knowledge in a society moving into a post-industrial period. In addition, there are economic factors which make science, and more particularly technology, an integral part of the development of western industrial society. The balance between the sometimes countervailing influences of these factors, in the period leading up to the moratorium, was complex, and a full analysis will not be attempted here. Rather, the following discussion seeks to clarify some of the more significant elements involved.

Science has been described as having a fairly cohesive normative structure which provides motivation and legitimation for the search for knowledge. This structure was initially outlined by Merton (1942), and has been labelled by Sklair (1973:43) the 'CUDOS syndrome': communalism, universalism, disinterestedness and organised scepticism. The results of science are thus supposed to be shared, impersonal, developed in accord with the notion of science for science's sake, and continually open to careful scrutiny by peers (Storer, 1966). This requires a cultural milieu free from overt political interference in order to promote an untrammelled, and objective, search for knowledge. However, it is recognised by many observers that such an ideal is rarely achieved in practice, and science is often characterised by elements of dogmatism and self-seeking. Indeed, Merton himself recognised the necessity of postulating a set of 'counter-norms' (Merton and Barber, 1963), and Mitroff (1974:592) has listed the following: faith in the moral virtue of irrationality as well as rationality; emotional commitment; particularism; solitariness; interestedness; and organised dogmatism. Mulkay (1979:72) has argued that the norms of science should be regarded as a 'repertoire of standardised verbal formulations' which can be flexibly deployed in arguments about the status of science, rather than as 'defining clear social obligations to which scientists generally conform'. However, many people, both scientists and non-scientists, feel that while the real world is morally imperfect, the ideal is still to be strived for, and attempts to undermine the neutrality of scientific advance should be resisted.

In the United States of America, this view has been strengthened by appeals to the first amendment to the Constitution, and the responsibility of government to safeguard the conduct of scientific research from unnecessary and unreasonable political intervention. Szybalski, an American scientist, quoted a US Supreme Court judgment of 1965 which underlines this point:

> The right of freedom of speech and press includes not only the right to utter or to print, but the right to distribute, the right to receive, the right to read . . . and *freedom of enquiry*, freedom of thought, and freedom to teach . . . indeed the freedom of the entire university community. (Szybalski, 1978:97, emphasis in original)

Szybalski went on to point out the moral concerning the dangers of what he called the 'politicization of science' which could only distract scientists from their 'primary goal of pursuing knowledge'. Another scientist, Medawar, referring to the DNA controversy itself, has argued that 'short of abolishing the scientific profession altogether no legislation can ever be enforced that will seriously impede the scientists' determination to come to a deeper understanding of the material world' (Medawar, 1977). Hence, while recognising the possibility of falling short of the ideal, many scientists, and a large proportion of the lay public, espouse an ideology of free enquiry in a free society based upon an ideal which emphasises objectivity and disinterestedness.

Several historical examples of the dangers of interfering with this ideological utopia are paraded before its critics. The most notable are Galileo and Lysenko, and the infamous Scopes 'Monkey Trial' concerning the teaching of evolutionary theory in schools. In a polemical mood, Szybalski (1978:127) has argued that it is as dangerous to base scientific laws on 'metaphysical, theological, and mystical grounds, or on some political doctrine' as it is to draw 'absurd analogies to science fiction . . . or a series of movies on the Frankenstein monster. . . .'

It is not, however, the case that such views are universally held. Both in the United States of America and in Britain this view of science as neutral and benign, and as needing to be nurtured in an open and free society, is seen by sections of the scientific community as both naive and potentially deleterious to development in the world of 'big science'. De Solla Price (1963) argued that the image of the gentleman scientist working quietly in his own laboratory has long been superseded by the reality of the industrial or university scientist employed by a large and well-staffed institution. The notion that, as Piel (1979:43) put it, 'A scientist can recognise no authority but his own judgement' can no longer be considered wholly appropriate in view of the funding constraints on the directions of scientific research, and of the broader social implications of science, especially when research may have potentially hazardous side effects. It is therefore not enough to simplistically link limits on scientific enquiry with limits on 'public freedom of citizenship' as Piel did, and it is necessary to recognise that scientific and technological advances do not occur in a political and social vacuum (see, for example, Rose and Rose, 1969).

Such a view, promulgated in the United States in the late 1930s by the American Association for Scientific Workers, and in Britain by Haldane and Bernal, has become increasingly accepted among many scientists. In the 1960s and 1970s, several groups of concerned scientists publicised the potential dangers of a wide range of scientific and technological advances, including, most notably, research into nuclear and biological weapons, which were consuming a disproportionate amount of the funds available for research both in Europe and America. As Krimsky notes:

> for several years prior to the advent of . . . recombinant DNA . . . techniques, there had been concerns about related research problems within the scientific community. These concerns, which were precursors to many or most of the elements in the later debate, reflected more general problems related to science's ethical, social, and political ramifications. (Krimsky, 1982:13)

Both he and Rose and Rose (1976) suggested that the Vietnam war provided the watershed for raising the consciousness of the working scientist (as well as the public) about the possible dangers in scientific and technological advance. Krimsky (1982:14) referred to the 'consequences of American military adventure abroad' which became the 'major influence on those who were later to be actors in the rDNA debate', as some saw participation in weapons research morally abhorrent, while others felt it would serve to reduce casualties and shorten the war.[1] In Britain, Rose and Rose (1976:14) saw the war as marking a break between the quiescence of the scientific conscience in the 1950s, and the development of Russell's International War Crimes Tribunal in 1966, the CND movement in 1967, and the British Society for Social Responsibility in Science (BSSRS) in 1968, which aimed at 'self-education for scientists concerning the control of science and expressing the abuses of science, with the goal of politicising and mobilising increasing numbers of scientists and science students' (Rose and Rose, 1976:19). Thus the ideological debate about the status and role of science and technology in contemporary society was well to the forefront of discussion among the informed public, both in Britain and America. It was reaching a climax at the same time as the potential hazards of using recombinant DNA techniques became an issue for scientific and public concern.

The significance of this rising tide of consciousness among scientists can be linked to a fundamental socio-economic change in post-war western industrialism, which highlights the rising importance of knowledge in the production process. Bell (1973:135) and Galbraith (1969:271) have suggested that, in the United States of America, professional and technical employees will soon form the second largest occupational group apart from clerks.[2] Increase in the proportion of these groups in society establishes a significant 'knowledge class' with an associated rise in intellectual technology to facilitate rational planning, prediction, monitoring and growth in society as a whole. Bell subsumed these changes under an axial principle of the development of theoretical as opposed to empirical knowledge, with a concomitant rise in institutions such as universities and research organisations to supersede the business firms of the previous industrial society. This reveals the development of the 'post-industrial society' which replaces the declining sectors of manufacturing industry with the rise of a new service economy (Bell, 1973:12–33).

On the face of it, there are other indicators to mark the changing emphasis noted by Bell, Galbraith and also Touraine (1971). De Solla Price (1963) calculated an exponential growth in the number of scientific journals and papers since 1750, and suggested that the number of scientists with PhDs may be increasing at a similar rate, especially since 1945. The proportion of the Gross National Product devoted to research and development expenditure has also increased dramatically in post-war years (Sklair, 1973), although, as Kumar (1976:457) noted, only a relatively small proportion is spent on basic research. Third, there is the recognisable rise in the importance of expert knowledge as the division of labour becomes so specialised that the lay public may become mystified when faced with coping with even the most commonplace technical problems of everyday life. Galbraith (1969) in fact argued that power is slowly being transferred to a 'technostructure' ruled by an expert 'technocracy' with a virtual monopoly of scientific and technical knowledge.

Other commentators such as Habermas have suggested that society is moving into a phase of late capitalism. This is characterised by a process of the demystification of the locus and workings of power in society. The 'hand of the state' becomes more clearly visible as the planning process becomes more accessible to a greater proportion of the population. Attention is drawn to the nature and application of social control in society, thereby endangering the hegemony of the state and its agencies. The result, according to Habermas (1976), is a 'crisis of legitimation' as more and more areas of social life are seen by the population at large as becoming politicised, and the state needs to counter the breakdown of support and loyalty to the existing order.

This 'crisis' is as much the result of changes in the basic economic structure of society as of the general growth of knowledge (Habermas, 1968) which has progressively demystified the apparent inevitability of some areas of technological development, especially through concerns with environmentalism. The popularity of books expressing concern with a wide range of environmental problems in the 1960s and early 1970s forms part of the backdrop to the moratorium on genetic manipulation research agreed at the Asilomar conference in 1974. Nelkin (1977b) has suggested that Rachel Carson's *The Silent Spring*, first published in America in 1962, marks the renewal of public interest in environmental pollution. Other works which followed, though with different remits, included Barry Commoner's *Science and Survival* (1968), G. Rattray Taylor's *The Biological Time Bomb* (1968), Paul Erlich's *The Population Bomb* (1968), Ward and Dubos's *Only One Earth* (1972), and Meadows *et al.*'s *The Limits to Growth* (1972). By the mid-1970s, however, these concerns had begun to make less impact on the public at large. Sandbach (1978) noted that, whereas in the 1960s and early 1970s, book and magazine sales were high, environmental groups legion and prophets of doom dominated the media, an analysis of news media and literature, public and opinion and social survey data, and environmental legislation in the mid-1970s shows a major decline in radical environmentalism, although traditional conservation and preservation groups have continued to flourish both in Britain and America.

An important element in the rise and fall of the 'doomsday syndrome' (Maddox, 1972) was the impact of new biological techniques, including recombinant DNA research, in the area of human reproduction and heredity. Popular literature has concerned itself with the latter area for centuries, with noteworthy contributions from Mary Shelley, H. G. Wells and Aldous Huxley, among others. More contemporary science fiction has used modern knowledge to underpin these 'folk devils' (Cohen, 1973) with the semblance of credibility, and 1969 saw the publication of one of the most celebrated examples, *The Andromeda Strain* by M. Crichton. As long ago as 1958, Walter Hirsch pointed out that, after conventional 'love interest', the effects of technology, basically of an unanticipated nature, formed the main social problem addressed by science fiction. He went on to state that the major themes are 'biological mutations which have been produced experimentally and then get out of hand' (Hirsch, 1962:266).

The specific concerns of science fiction writers in the late 1960s and early 1970s were complemented to some extent by development in the media as a whole. Thus *Time* magazine had a 1971 headline 'The body: from baby hatcheries to "xeroxing" human beings' over an article which began with a reasonable description of some of the problems associated with genetic counselling, including amniocentesis and artificial insemination, and went on to foreshadowing the development of 'test-tube' babies. However, the article then moved from this to resurrect the folk devils of human cloning and genetic surgery:

> Prophylaxis is important, but man's molecular manipulations need hardly be confined to the prevention and cure of disease. His understanding of the mechanisms of life opens the door to genetic engineering and control of the very process of evolution . . . *Soon, man will be able to create man – and even superman.* (*Time*, 1971:27, emphasis added)

Geneticists and molecular biologists were not unaware of these broader public concerns, and while recognising possible dangers, some attempted to allay them in print. For example, Bernard Davis, writing in *Nature*, noted that while 'Promethean predictions of unlimited control . . . have led the public to expect the blueprinting of human personalities', most scientists, and especially geneticists, 'have had more restrained second thoughts' (Davis, 1970:1279). He then went on to argue that any small dangers that exist are social, not scientific problems, in that they would only form part of a much larger number of already present dangers which society already satisfactorily confronts. The period leading up to Asilomar and the 'Berg letter' saw a heightened awareness, both by scientists and the public at large, of the potential dangers stemming not only from the environmental 'crisis' brought on by advances in technology, but also from the possible misuse of technology in the field of human genetics. Impressionalistically, much of the public debate seems to have been stylised and stereo-typical, drawing its images from 'folk memory', and its horrors from *Frankenstein* or *Brave New World*.[3] Expert debate was characterised by disagreement about the extent of the dangers and the proximity of the apocalypse. It is clear that, as noted by Douglas (1975), societal values and interests were seen to be under general threat. In such conditions, certain encroachments can become defined as polluting the moral environment of society, and can lead to a strong social reaction.[4] Society is at once threat-ened, and made more cohesive by recognising such threats, and by responding to them in a process described by Cohen as a 'moral panic', where 'sometimes the object of the panic is quite novel and at other times it is something which has been in existence long enough, but suddenly appears in the limelight' (Cohen, 1973:9).

Genetic manipulation was a novel technique which breathed new life into long-standing fears. The call for the moratorium, and the subsequent generation of hazard scenarios by some leading scientists, was seen to indicate a threat not only because of poss-ible biological pollution by new disease entities, but also through the potential pollution of the moral environment by scientists 'playing God', altering genetic endowment, and transgressing the natural boundaries between non-interbreeding species.

It is, of course, impossible to accurately partition the reaction into specific worries about the possible hazards of genetic manipu-lation techniques on the one hand, and the crystallisation of more general fears about the wider social implications of scientific advances on the other. What can be said, and has already been illustrated, is that as the debate progressed such a distinction became part of the rhetoric of, and actions in, the controversy. There was a broad move by some scientists towards depoliticising the debate by narrowing it to the technicalities of avoiding poten-tial hazards. Others argued for a more inclusive view in which the possible wider social implications would be paramount. The introduction of social policies in the form of agencies to regulate genetic manipulation experiments thus represents a compromise, and can be seen as a coping strategy to both satisfy the requirements of scientists to perform experiments, and to reassure the wider society that their fears would be successfully ameliorated, and the dangers controlled.

There can be no doubt that a significant set of factors in the establishment of the regulatory compromise was the recognition that there were important industrial, medical and agricultural benefits to be gained from the eventual application of genetic manipulation techniques. Wright (1978:1425 *passim*), for example, has argued that 'economic and competitive pressures were influential in shaping needs to exploit genetic manipulation on both sides of the Atlantic' (see also Yoxen, 1983). The particular agency through which the link between industrial production and the new techniques was expressed should be noted. Large (multinational) corporations involved in the production of chemicals and drugs were initially slow in implementing the possible benefits of genetic manipulation. A combination of individual scientist entrepreneurs and venture capital has made most of the running, and much of the pressure from industry for the relaxation of regulations has come from this quarter rather than the large corporations. It is, of course, dangerous to project this tendency back to the period before the moratorium, but it is at least consistent with other evidence that it was mainly academic scientists who articulated the potential industrial applications.[5]

The practice of public participation

Events following the moratorium on recombinant DNA research and the discussions which took place in Britain have been covered in earlier chapters.[6] Scientific uncertainty and wider social perspectives were gradually replaced by an increasing focus on the decision making machinery necessary to make assessments of possible hazards and appropriate precautions. Scientists had been largely successful in defining the situation as requiring some form of risk assessment. The novelty of GMAG was that it formally involved a number of individuals who did not possess the relevant scientific expertise in the process of technical decision making. This section first outlines some of the general issues relating to the interaction of expert and non-expert on technical matters, and then examines the role of the non-expert representatives of the public interest on GMAG.

Much decision making in modern society places a greater reliance upon those with technical knowledge and expertise. Experts have always been present in society in the form of individuals with some monopoly over esoteric knowledge or skill, as, for example, in the case of religious functionaries in feudal times. But the rapid development of knowledge and the associated burgeoning of education has meant not only that experts have begun to proliferate, but that the circumstances under which they may disagree have also multiplied.

Over a decade ago, Mazur highlighted disagreement between scientific experts by comparing different evaluations of two distinct controversies: radiation levels and fluoridation (Mazur, 1973). He characterised the disputes in these two areas as empty, and showed how rhetorical devices such as 'There is no evidence to show that . . .' were used to deny the validity of opposing arguments. He concluded that many disputes between experts were not disputes about conclusions but about premises, and that, in the choice of ground from which to attack, 'Experts tend to behave as other persons behave when they engage in a controversy' (Mazur, 1973:259). Thus, as Fischoff *et al.* (1981) have stated, it may be better to think of the 'opinions of experts' rather than 'expert opinion'. One by-product of such disputes is that they reveal otherwise hidden arguments and assumptions, thereby opening up the debate to outsiders, and creating the possibility of intervention by non-experts.

Attempts to discover either the basis upon which expert judgment is made, or the premises from which it stems, are not, of course, limited to clashes between experts.[7] Lay individuals and groups in the wider society increasingly feel the need to challenge those who appear to monopolise knowledge, as evidenced by the controversies over the relationship between scientific and technological advances and public policy. As Nelkin has pointed out, those proposing innovations

> define the decision and the issues involved primarily as technical – subject to objective criteria. . . . Opposition groups, on the other hand, perceive such decisions in a political light. . . . They try to show that important questions involve political choices and that these can be obscured by technical criteria. In the end, they seek a role in making social choices. (Nelkin, 1979:17)

As has been shown, the controversy over genetic manipulation exemplifies this general pattern. The wider social issues and political choices were in practice settled by the early deliberations of the Ashby Working Party and at the Asilomar conference.

The relationship between technological advance and public policy can, according to Nelkin (1977a), be conceptualised in terms of broad societal control. She suggested that three main areas exist: participatory, reactive, and anticipatory, and that each could form the basis for disputes involving expert knowledge. She placed the involvement of citizens' groups, and scientists in political activity, into the first area. Through this involvement, attempts are made to bring wider social or community values to bear on technological decisions. Nelkin (1977b:418) noted that citizen group activity can be cumbersome and subject to disproportionate influence by interest groups, especially when concerned with complex and controversial technological decisions. They also pose special problems for the non-expert since, often, technical data can only be challenged by a counter-expert.

Those institutions in society which do not actively seek out implications, but deal with already existing problems, are reactive: 'They register complaints, but do not try to discern future impacts of nascent technologies' (Nelkin, 1977b:421). Such controls tend to be legal or administrative. Regulatory agencies, such as independent commissions, oversee industrial practices by setting rules and safety standards, and issuing permits. For example, the Health and Safety Executive in Britain performs these kinds of functions. In some cases, for example in the Atomic Energy Commission, the controls extend beyond such matters to determining prices and profits, and entry to the market. Finally, anticipatory controls are based upon the planning of future research in order to discover the future impacts of newly emerging technologies. This is often accomplished by some process of technology assessment, which is itself open to severe criticism and may only broaden the basis for dispute. Nelkin (1979) clearly recognised the political nature of expertise in all three areas, as scientists are called in to buttress a wide range of political positions. Wynne (1982) also recognised this, while emphasising the need for ritual and ceremonial illustrated for him by the Windscale Inquiry, as a way of disguising some of these political decisions:

> Although the Inquiry propagated belief in the democratic control of technology, it did so with a fervour which betrayed the cultivation of the myth of free control at the very point where it was most contradicted by reality. (Wynne, 1982:10)

In addition to their practical function, arguments about technical knowledge and the basis of expert judgment can take on symbolic

163

significance in the process of legitimation in the wider society (Habermas, 1976). The need to give comfort and reassurance in face of the crucial uncertainties of decision making and policy analysis on the frontiers of knowledge may result in the formation of what may be called an 'ideology of expertise' which performs just this symbolic function. Societies develop appropriate rituals and ceremonies so that a complete halt to progress is averted. Thus GMAG had both practical and symbolic functions although, as will become clear from the discussion, the symbolic nature of the membership of GMAG was somewhat less than might be thought for a committee of this type. For example, the scientific members seem to have been chosen as much on the basis of their knowledge of relevant areas as for their scientific eminence.

As has already been noted, in making recommendations on the composition of GMAG the Williams Working Party did not have any debate about what constituted an appropriate number of representatives of the public interest. Nor was their relationship to any constituency of the general public ever clearly defined. It was the Minister, Shirley Williams, who appointed individuals whom she felt would be able to make an intelligent appraisal of the sophisticated scientific arguments, and would be able to keep a watching brief over the process of vetting applications for research projects by scientists. It was not, therefore, necessarily desirable to have a scientific background, although an interest in science or public health was clearly an advantage. Indeed, Shirley Williams, in evidence to the Select Committee (1979:155 para. 642), stated that public interest representatives were deliberately chosen because they are not themselves working now in the field, because they are not beholden to either industry or to a research institute'. Thus, lack of technical expertise in the area was intended to be symbolic of neutrality and independence of the interests of science or industry. This, of course, raises a problem, since individuals who met the criterion of independence would be likely to find it difficult or impossible to take a full part in any technical decision making.

It is clear that individuals were chosen for quite particularistic reasons, perhaps from a group known to be available for this kind of public appointment in Whitehall, which is not to say that those appointed were not well qualified for the task. Certainly, one representative felt that she had been picked because of friendship with Shirley Williams, while another, somewhat disingenuously, was very flattered to be chosen 'out of the blue'.

The need to represent the legitimate public concerns about the potentially hazardous nature of genetic manipulation research was conditioned by a number of independent influences in addition to the general factors noted earlier in the chapter. In the USA, the principle had been established through representation to the Director of the NIH, and the formation of a series of 'citizens' courts'.[8] The best-known of these was the Cambridge Experimentation Review Board, which was instituted by the Mayor of Cambridge, Mass., to judge whether or not Harvard University was to be allowed to continue research in this field (Krimsky, 1978 and 1982: ch. 22). In Britain, a second factor was the public concern expressed in 1973 about the two deaths at the London School of Hygiene and Tropical Medicine from a virus which had escaped from a supposedly safe laboratory.[9] This led to a revival of fears of widespread infection of the population by a terrible disease, released either by accident or oversight, and resulting in death and destruction. The event was widely publicised in the media, and questions were asked in the House of Commons. A third factor was the formation of a powerful interest group led by the white collar union, ASTMS, and including several other unions with members employed in scientific laboratories, which lobbied the Labour Government for more effective health and safety measures for their members, and for the inclusion of trade union representatives on any body to oversee genetic manipulation. Finally, Fred Mulley and Shirley Williams, the two Ministers involved in setting up GMAG, were also personally committed to providing as wide a basis as possible for decision making about potential hazards within the broad rubric of Labour Party policy on the participation of the public in such areas of concern.

However, it was never made clear just how the lay members of GMAG were ever going to be able to perform the role of representing the public interest, thereby highlighting the question of whether their participation was to fulfil a symbolic rather than a practical function. Nelkin (1978) discussed the problem of including members of the lay public and practising scientists on the same committees, focusing on the crucial question of ultimate responsibility for decisions about controversial scientific and technical questions. She concluded, referring to American human experimentation review boards, that 'A survey of these institutional review boards found substantive differences between the perspective of the laymen and the scientists who were participating' (Nelkin, 1978:203). This difference in approach is related to another problem faced by mixed review boards, namely the perceived differences in status between the two groups of participants. The nature of many of the scientific questions necessarily gives the scientists a general advantage in any technical discussions with ordinary citizens, however well-informed. Very often, the scientists themselves are leading experts in the field in question, and able to be authoritative on that dimension. Krimsky, in his discussion of the lay Cambridge, Mass. 'citizens' court', where he was a participant, concluded that scientists and citizens do not mix well when big scientific and technical decisions need to be made. He stated that 'There is a strong tendency for the scientists to play a dominant and elitist role that intimidates lay persons' (Krimsky, 1978:42).

These kinds of considerations do not appear to have been seriously taken into account during the development of the Genetic Manipulation Advisory Group in Britain.[10] The composition of GMAG was so structured that scientists were the largest single group, thus emphasising the importance of technical knowledge. However, it must also be noted that, by the normal standards of such committees, scientists might well have been expected to have had an absolute majority over other groups. The thinking behind British initiatives in this area is probably best summarised by an editorial in *Nature* which stated:

> What has to be done is that all sides must see that their case is *at least taken into account* by whoever makes the decisions, and they must be satisfied that the choice of the decision-maker and the form of the enquiry does not prejudge the case. (Anon, 1979a:675, emphasis added)

Thus the role of the citizen was more that of an umpire or referee than a real participant in the decision making process. One of the chairmen of GMAG described the role of the public interest representatives as important 'neutrals' in clashes between the scientists and the trade union representatives. The representatives of the public interest were not really there to make hard, scientific and technical decisions, but rather to ensure that all points of view, especially those of the public at large, were 'taken into account'. One of them underlined this 'outsider' status when she said, during the course of an interview, that 'GMAG has taken account of our concerns when we've had any'. When asked about the effectiveness of the contributions made by her colleagues and herself, she said that they were 'probably the least influential' in the decision making process. A second public interest representative explained:

> 'What we can do is see that the methods being used by GMAG to investigate and check new experiments and so on, seem

from the general, educated point of view, to be efficient and adequate. . . . But, of course, we couldn't tell whether they were adequate if somebody were really trying to get around them. It would only be the scientist who could catch the scientist. But then, we're there to support whichever side we feel, in any dispute, has the greater logic and sense.'

A third representative talked of 'justice being seen to be done', while a fourth expressed frustration because

'as a lay representative on the committee, I was judging in terms, to put it most crudely, of how impressed I had been with one or another scientist. . . . You know, it raised in my mind the whole issue of the inherent difficulties of public participation in technical decision making, because, you know, if they had set up this committee without public interest representatives, I would have been the first to protest.'

An issue closely related to the ability of public interest representatives to effectively participate as full committee members was the kind of training provided to familiarise them with the basic knowledge required to understand the techniques of recombinant DNA research. In the United States, scientists seem to have been much readier than in Britain to recognise the esoteric nature of their scientific and technical knowledge.[11] This appears to have been the case in the setting up of the all-lay Cambridge Experimentation Review Board (CERB). Among other things, members of CERB were

(a) given a file of technical documents, including NIH guidelines, articles from *Science*, *Nature*, etc.;
(b) given the use of a technical assistant, with training in the biological sciences, both for help with difficult concepts, and to keep them up-to-date with the developments in the literature; and
(c) allowed to witness a 'dummy' experiment taking place in a proper laboratory. (Krimsky, 1978:39–40)

The situation with GMAG at its inception was rather different. According to one of the public interest representatives, Professor Mark Richmond gave a 'beautiful' illustrated lecture outlining the recombinant DNA techniques. Another remembers there being, in addition to a lecture for 'non-experts', a 'whole lot of reading material', but then goes on to say that 'anybody like me with a reasonable intelligence could not understand all the material without devoting at least 6 months of 8 hours study every day to it'. In fact, it was for precisely this reason that Gibson, the Secretary of GMAG, described the early sessions as purely educational, and commented on the difficulty of recruiting public interest representatives at all.[12]

No real attempt was made to instruct the public interest representatives in laboratory techniques. The only occasions when experience of laboratory life was available to them was on visits to check whether appropriate containment levels were up to the standards laid down by GMAG prior to the approval of certain research projects. Such visits were never frequent, and diminished in number over the life of GMAG. Hence the public interest representatives who were relatively unfamiliar with the techniques of molecular biology in the laboratory were reliant upon the greater expertise of the practising scientists in this area of judgment.

One of the participants, who was not a professional scientist, welcomed the site visits as providing 'greater insight into what was happening' and found she was able to make certain practical suggestions which were accepted as minor modifications in laboratory practice. However, she also clearly felt that her initiatives related primarily to non-scientific areas, and she left it to the scientists in the visiting parties to concentrate on more technical matters.

Another public interest representative admitted that, in the course of committee meetings, it was sometimes difficult for non-scientists to make useful contributions because of the 'funny looks' that naive questions occasionally brought. Her justification for continuing this practice lay in her feeling that, as a representative of a largely ignorant public, it was her job to question anything with which she was unfamiliar. Her comment on the knowledge gained from site visits was 'I would never have known it otherwise'. Site visits performed the dual function of allowing members of the committee to inspect facilities and meet researchers, and of informing and instructing those committee members who were less familiar with actual laboratory practice. This was rather different from the CERB experience, where members were carefully taken through a model experiment in a laboratory step by step.

In terms of active participation in decision making, and the ability to influence outcomes, all the public interest representatives on GMAG who were interviewed, with the exception of Maddox, felt a little out of their depth with the more specialised scientific and technical detail upon which they had to exercise their judgment, although it was clear that they could always direct questions to their more knowledgeable colleagues on the committee.[13] However, as one of them noted, occasionally these colleagues 'told me where I got off!'.

In the particular context of active participation, most suggested that the only really effective public interest representative was Maddox, who, while not being a practising scientist, was both competent and knowledgeable enough to play an equal part in the scientific and technical discussions between the scientific members of the committee. Evidence of this can be found in his redraft of Brenner's theoretical categorisation of hazards into a practicable scheme.[14] Most of the other public interest representatives had but limited success in mainly peripheral matters. They were able to voice opinions, if not influence decisions, when policy issues arose. It comes as no surprise, therefore, to gather that some, at least, found their experiences frustrating, and their impact on specific decisions limited. The notable exception was also the individual with the greatest expertise, and the highest status in the scientific community through prior editorship of a major, international, scientific journal.

Thus the possession of technical knowledge was crucial to full participation in many areas of decision making on GMAG. Given that one of the important criteria in the selection of public interest representatives tended to exclude those with such knowledge, their reservations and difficulties are understandable. Further, the possibility of ameliorating or coping with the effects of this condition was prevented by the fact that the activities of GMAG were conducted behind closed doors, and effectively screened from the public at large by the Official Secrets Act.

The effect of this can be illustrated by a brief comparison with the operation of the American Recombinant DNA Advisory Committee (RAC). In 1979 its membership was approximately doubled to twenty-five, and contained one-third public representation, including several prominent critics of existing policies (Krimsky, 1982:156). Much of the discussion was open to the public; extensive reports of meetings were available, including details of arguments for and against. These arrangements made possible informed comment by outsiders with, potentially at least, two consequences. First, such extension of the debate makes available a much wider pool of arguments and opinions. This pool can be used as a resource in discussions and arguments with experts. Second, those outside the decision making arena who become engaged in the extended debate can form a 'constituency'.[15]

Both of these possibilities were denied to the public interest representatives on GMAG. As far as a pool of arguments was concerned, they were effectively left to their own resources. Formally, members of GMAG were individuals who represented an interest, rather than acting as delegates of particular interest groups. In effect, however, the scientists, industrialists and trade unionists all had constituencies to which they could turn if necessary, as also did the civil servants.[16] Thus while, for example, the TUC representatives were always able to threaten a return to their constituency as a way of ensuring that their point of view was included in any discussion, no such option appeared to be available to the public interest representatives. Their ultimate sanction would have been to have 'gone public'. This was potentially a very powerful sanction, but its strength was its limitation. The application of pressure in a series of negotiations requires a graded set of threats; to have 'gone public' might well have called the credibility or even the existence of GMAG into doubt. A limited number of intermediate responses were, however, available. Most representatives felt that a word with the chairman would be sufficient in most cases. One did suggest that 'Had there been any hanky-panky. . . . I would have reported that to Mrs Williams'.

When the question of constituency responsibility was raised in interviews, most of the public interest representatives saw themselves as individuals, rather than as members of a group together with other representatives. They rarely met outside the confines of the GMAG meetings, and never appeared to form the sort of caucus characteristic of the trade unionists. As one public interest representative put it, 'One can't go back to anybody and say "What line should we take on this?" ', while another stated quite clearly 'We do not have delegates; we're individuals'. This public interest representative went on to distinguish between a constituency model and a 'British clubby way', with GMAG placed firmly in the latter tradition as far as the public interest representatives were concerned.

In contrast to the constituency model, with its accountable membership whose function is to ensure that interests are taken into account, and to receive from and feed back to those who are being represented, GMAG was composed of individual members chosen for their personal qualities who, in the case of the public interest representatives, somehow ensured that the public concerns were heard. Their function was to help create and monitor a balance of views. Hence, while other representatives needed to be technically knowledgeable so that they could effectively represent scientific and union interests, what was required of public interest representatives was the combination of the quali-

ties of independent judgment and clarity of mind. While, in principle, this is an admirable arrangement, it is apparent from the
research findings, that, in practice, it was very difficult for the
public interest representatives to effectively play their role. The
main reasons for this were their relative ignorance of the scientific
and technical details, their inferior status *vis-à-vis* the practising
scientists on the committee, and the lack of real sanctions should
things not be seen to go well.[17]

It is clear, then, that their relative lack of ability to take an
active part in the practicalities of decision making was a by-
product of the neutrality of the public interest representatives,
and a consequence of the organisational structure of GMAG. A
further factor was that, as scientific concern about the likelihood
of real hazards lessened, there were correspondingly fewer differences of opinion between experts, and hence fewer points of entry
into the debate for lay persons.

In turning to the less technical areas of decision making, to the
more symbolic aspects of the public interest role, and in assessing
the success of the experiment in public policy making, a more
positive evaluation is justified. It has been shown that the range
of different interpretations of their role given by the public interest
representatives themselves was not very large. One, for example,
saw the job as providing a 'safety net' to check on whether scientists were 'clearly going outside the immediate needs of society,
or research for the future, into some dangerous area'. Another
saw the job as ensuring that 'justice was being done' to the many
arguments on all sides of the debate as part of the normal system
of 'checks and balances', and playing the role of juror.[18] As noted
earlier, such symbolic functions are important for society. Indeed,
all the public interest representatives and many of the other
members of GMAG were strongly in favour of public representation, and cited examples of its benefits. One frequent comment
was that the presence of non-experts was a constant reminder of
the need to take account of public opinion. Whilst it might be
difficult to cite examples of their influence on specific decisions,
it was argued that the public interest representatives had a more
diffuse effect on the general tenor of GMAG's considerations and
decisions. The presence of neutrals in disputes between other
groups, for example over the confidentiality issue, had helped to
prevent excessive polarisation. It was also suggested that from the
point of view of scientists and the government, the existence of
public representation deflected possible criticism that scientists
were being allowed to regulate themselves.

In conclusion, the major limitations of this experiment in lay involvement in the formation of science policy devolve in large part from the structural features of GMAG. These might have been ameliorated to some extent if a more open and accessible mode of operation had been adopted, or if the resources for dealing with technical matters had been made available to the public interest representatives. However, the limitations are much more than just a local difficulty; they are symptomatic of the real and deep problems of representing the public interest, or involving non-experts in technical decision making. The GMAG experience did not solve these problems, but it none the less represents considerable progress.

7
Conclusion

Since the initial upsurge of concern about genetic manipulation in the mid-1970s there has been a continued, but uneven, decline, first in the scientific and then in the public belief in the likelihood of any risks. At the point when GMAG began its operations in December 1976, the change in scientific opinion was already well under way, although public apprehension was still high. By 1984 both had diminished to well beyond the point where GMAG could be terminated and its occupational health responsibilities transferred to the new Advisory Committee on Genetic Manipulation under the HSE.

It has often been remarked that GMAG was a novel experiment in 'social democracy in science', because of its inclusion of interest groups, particularly representatives of the public interest. So, how well has the experiment worked?

It should first be re-emphasised that genetic manipulation is a research technique. Although it first appeared within the broad research area of molecular biology it was clearly recognised that its application would soon spread to many areas of biological and medical research, and that many users would be unfamiliar in dealing with potentially pathogenic living organisms. Thus the regulatory agencies were concerned with controlling a technology rather than the content of an area of science *per se*.

More locally, any assessment of the success of GMAG must recognise the particular circumstances of, and constraints on, the British response to the worries over genetic manipulation. These included the concentration of funding of research (and hence its control) in the United Kingdom, the existence of a powerful union representing many of the workers in the field and the Health and Safety at Work Act which potentially covered research in the area. The British experience was characterised by relative calm and the absence of the open and public controversy found in the United States. The issue of 'academic freedom' was not raised in public in Britain; environmentalist groups remained quiet, and from the beginning there appeared a consensus amongst scientists, administrators and trade unionists that differing degrees of (possible) risk were involved, and that some form of regulatory machinery should be created to grade risks and precautions.

But, despite these significant structural influences on the establishment of GMAG as the regulatory agency, and the very important constraint of comparability with the American guidelines, one enduring impression of the character of the British response is the relatively small and exclusive circle from which the scientific membership of the various committees was drawn – the upper reaches of the biomedical establishment. Moreover, the concentration in this book upon events inside the working parties and GMAG is more than just a consequence of the type of research and the mode of presentation adopted. A veil of confidentiality surrounded the work of the Group; despite the occasional leak, there was no publication of, and hence no external discussion of, its deliberations.[1] Thus GMAG was rather inward-looking. To be sure, this is not meant to imply that GMAG was in any strong sense cut off from the events of the outside world, but rather that, from the point of view of outsiders, events impinged on the Group as on a 'black box'. Decisions emerged, but, except in discussion with individual scientists, the Group did not give reasons or explanations.[2]

This containment of the arena of discussion was effectively aided by the failure, with one exception, of the Royal Society as a body to take any active part in debate. As a related point, the only important public meeting at which the issues surrounding genetic manipulation have been discussed was that organised by ASTMS, in October 1978. In short, however sound the decisions of the Group, and however well the public interest in particular was represented, the novel aspects of the GMAG experiment did not include a departure from the British tradition of decision making behind closed doors.[3]

That a more open stance would have made any substantial difference to GMAG's policy decisions – for example, in the rate of relaxation of containment levels – is arguable. Opening general policy discussions to public view would have led to a more informed public awareness, but, equally, might have encouraged interest group representatives to merely rehearse the positions they represented and have caused delay, rather than leading to engagement in realistic negotiations. However, it should not have been beyond the wit of GMAG to have produced some suitably drafted record of its meetings to aid greater public awareness and participation.

Turning to more concrete matters, the effectiveness of the Group's regime depended vitally on what actually happened in laboratories, and on the actions of the local genetic manipulation safety committees. GMAG in general, and the trade unions in particular, placed great emphasis on the safety committees as the 'front line' of regulatory activity. Detailed knowledge about local laboratory conditions and practices was necessary for the implementation of the policies and categorisations advised by the Group. An effective committee, representative of all levels of staff involved, was required to ensure both compliance and confidence.

GMAG felt that the responsibilities placed on local safety committees were thoroughly and well discharged. The introduction of the new risk assessment scheme in 1979 endorsed this confidence by transferring a good deal of decision making power about categorisation to the local committees. As far as can be gathered, full and open discussion was the order of the day, and most felt that safety committees had performed a valuable function. There was a widespread feeling amongst all concerned with the area that a particularly successful effect had been the raising of the level of safety-consciousness in laboratories.[4]

Critics of the regulation of genetic manipulation have argued that it had an inhibiting effect on the development of research, and delayed the emergence of industrial biotechnology. GMAG itself has occasionally been charged with producing overstringent requirements leading to needless expenditure on high containment facilities, and a competitive disadvantage for British scientists. On the more general criticism of the inhibition of research, there was some initial effect, especially on the more important laboratories in the field. But however acutely this may have been felt by the scientists involved, it would seem to have been a relatively small price to pay, and one necessary for the research to be allowed to go ahead at all. There is almost no evidence of delays in the creation of a biotechnology industry which can be laid at the door of regulatory agencies. As far as Britain is concerned, any limitations on industrial progress have been inherent in the processes of setting up new industrial ventures, and were unaffected by regulation as such. GMAG made a point of dealing as speedily as possible with proposals from industry, and it was suggested in interviews that one multinational company located its research facility in this country precisely because of the stable regulatory climate.

In terms of the specific charges against the Group, it must be accepted that several high containment research facilities were built which turned out to be largely redundant almost as soon as they were completed, because of intervening relaxation of containment requirements. Although comparison is very difficult because of the different philosophies behind the two systems, it would seem that, overall, GMAG's containment requirements were probably just slightly higher than those of the NIH, and hence that there was a small competitive disadvantage. However, this had little general effect, but was significant at certain periods, and in the case of certain experiments. For the sake of balance it should be noted that in the early days there were a few cases of American researchers coming to Britain to perform experiments which were prohibited by the NIH regulations.

There were, then, some costs to science in general, and to British scientists in particular, as a result of the regulation on genetic manipulation.[5] There were also, of course, benefits. One unanticipated consequence of GMAG's activities was that, for example, scientists unfamiliar with genetic manipulation research could draw on the Group's list of approved host/vector systems in designing their experiments. However, the main benefit to be set against any negative cost must be that of public confidence in the visibly safe use of genetic manipulation techniques. The price put upon public reassurance determines the outcome of any such cost/benefit comparison.

The inclusion of representatives of the public interest as well as those of employers and employees was a brave attempt to broaden and democratise the decision making process in science, providing a unique exemplar for the development of science policy machinery, not only in Britain but elsewhere. It must be judged a qualified success. Early in the debate deep questions about the relationship between society and technical progress were raised. The Ashby report (1975:3) had asked 'how can the social values of the community at large be incorporated into decisions on science policy?', whilst simultaneously beginning the process whereby these issues became focused on to issues of health and safety. The original question still awaits a satisfactory answer, but the GMAG experiment has provided some useful pointers.

The lay representatives of the public interest were able to

176

monitor the activities of specialists with varying degrees of effectiveness. Their influence was never very great in any matter of real technical importance, with the possible exception of the change in the basis for assessing risk. Certainly, with their presence once GMAG was established, the wider society felt safer and better informed. Concern in Britain over potential hazards never reached the levels generated in America. A major lesson to be learned from GMAG is the need for, and usefulness of, the institutionalisation of a broad, non-specialist input into scientific and technological development. Further lessons are the need for a less secretive approach, and the provision of some form of independent scientific advice for non-specialists. But, even without these latter conditions, GMAG has shown that a non-specialist contribution is a viable proposition. However, GMAG's transformation into the ACGM under the HSE in 1984, lacking this broadly based membership, shows how public interest representation could be dispensed with at a decent interval after public concern had been assuaged.

The development and operation of a process for regulating genetic manipulation research is a contemporary example of the way in which both science and the wider society have confronted a controversy about potentially hazardous advances in scientific knowledge and its technological application. With hindsight, GMAG is viewed as having been largely successful, both in allaying public fears and overseeing the development of genetic manipulation in Britain. A more cynical view has it that GMAG worked well because the risks were less than first thought and failed to materialise. Whilst the story may have been quite different had the hazards eventuated, the fact that GMAG in its earlier years managed to draw together its disparate membership and to carry out its set tasks on the basis of little fact and much rhetoric in the face of opposing scientific and public concerns, and to do so with such conspicuous success, is the main surprise.

Notes

Chapter 1 Introduction

1 The main works are by Wright (1978), Krimsky (1982) and Hanson (1983). There also exists an Oral History Archive on the genetic manipulation controversy at the Massachusetts Institute of Technology. Several books detailing events in America, with very little commentary, have also been published, for example, Watson and Tooze (1981).

2 Apart from Wright (see note 1), the main commentators are Ravetz (1979) and Yoxen (1979a), although comments have come from groups such as the British Society for Social Responsibility in Science (1978 and 1979) and Science for the People (see Cockburn, 1981 and Yoxen, 1979b).

3 More details can be found in Appendix IV, 'A note on methods' in Bennett, Glasner and Travis (1984).

4 An important work in the development of the new approach was Barnes (1974). For a discussion of the general sociological position, including its history and current prospects, see, *inter alia*, Collins (1981) and Mulkay (1979). A collection of articles can be found in Barnes and Edge (1982). For a discussion of relativism and its critics, see Hollis and Lukes (1982).

5 Several books showing the disparate nature of the new approaches based on detailed empirical work include Edge and Mulkay (1976), Latour and Woolgar (1979), Collins and Pinch (1982) and Gilbert and Mulkay (1984).

Chapter 2 The development of a British policy

1 The NIH is the main governmental funding body for biomedical research in the USA.
2 These points are dealt with at greater length in Rogers (1977).
3 The 'flavour' of the debate is well described in Rogers (1977) and Wade (1977).
4 See Subcommittee on Science, Research and Technology, US House of Representatives, 94th Congress (1976); DHEW (NIH) (1976).
5 Quoted from the pre-conference literature for participants. Importantly, and quite explicitly, participants were chosen also on the basis of their ability to influence decision making machinery in their own countries.
6 In more sociological terms, estimation would have to be based on scientists' 'tacit knowledge' (Polanyi, 1958).
7 This is especially significant as drug resistance might otherwise be used as a 'genetic marker', or to select from the many clones produced in 'shotgun' experiments.
8 The 'ten-litre barrier' was a more or less arbitrary cut-off point to prevent large quantities of material being produced. Obviously, the smaller the amount of culture, the less the (possible) risk.
9 In practice, little direct 'risk assessment' research was performed, partly because scientists' estimates of the likely risks continued to fall.
10 Local legislation was proposed by several cities; this led, for example, to particular restrictions being placed on research in Cambridge (Mass.).
11 Subak-Sharpe had learned of the 'Berg letter' from one of its signatories, Baltimore, at a conference in Cold Spring Harbor. A number of other European scientists were also at the meeting and, feeling that a European policy was required, they wrote to John Kendrew, the General Secretary of the European Molecular Biology Organisation (EMBO).
12 The Agricultural Research Council was not sponsoring any relevant research (this would normally be funded by the MRC), and does not figure in participants' accounts of the development.
13 Such a distinction was made in 1975. See the discussion of the 'Oxford conference' below.
14 The actual level of stringency is not of direct relevance here – the concern is only with the perceptions of the proposed precautions.
15 They included Subak-Sharpe and Kornberg. One of the two secretaries was Vickers who had organised the MRC meeting on the implications of the interim Godber report.
16 There was satisfaction in some quarters at having ensured that in the terms of reference, and throughout the report, 'potential benefit' is placed before 'potential hazards'.
17 Wright (1978:1394) argues that the Working Party also went 'substantially beyond its charge'. Its general conclusion was that

suspension of the work could 'be no more than a pause, because the techniques open up exciting prospects for science and for its applications to society, and evidence we have received indicates that the potential hazards can be kept under control'.

18 This is not to imply that this was a consequence of Brenner's evidence.

19 Subak-Sharpe (personal comment, 1984) reported that, against the wishes of some of the organisers, a 'vote' was taken on whether or not it was felt that any real danger existed. Apparently the result was more or less evenly divided.

20 A fuller discussion of these points is to be found in Collins (1975) and Travis (1981).

21 Commenting on the ASTMS position at the time of the Williams Working Party, Wright (1978:1427) makes no mention of the first and fifth points.

22 ASTMS, with its long-term aim of increasing its membership amongst scientists, regarded itself as the rightful representative of employees' interests in the area, a view opposed by the Association of University Teachers (AUT) and the Committee of Vice-Chancellors and Principals (CVCP).

23 This is, of course, a point about the mechanisms of safety advice; the concern for safety as such cuts across this analytical distinction.

24 Once again, many of the following points are close to what later became official policy.

25 ASTMS also had the wider and longer-term goal of trade unions increasing their involvement in policy making about science and technology.

26 Of course, the Act dealt only with humans and the conceivable hazards were much more widely based.

27 It was by no means necessary, or even likely, that the HSE would take any action. There were, after all, many other areas of responsibility to claim its attention. On the other hand, it was legitimately open to trade unions (amongst others) to press particular cases. The following quotations, though published somewhat later, undoubtedly illustrate the general ASTMS approach at the time of the Williams Working Party.

> [The HASAW Act] is by no means a perfect piece of legislation because it is woolly and infested with the delaying words 'so far as is reasonably practicable' which have a history of emasculating good Parliamentary intentions. (Jenkins and Sherman, 1977:117)

> [The HASAW Act] provides a framework for trade union activities and has to be viewed as enabling legislation which will only be effective as those involved make it. . . . This will all need expertise and unions are finding it necessary to appoint full-time safety officers. Once again, the range and scope of collective

bargaining is being increased. (Jenkins and Sherman, 1979:106–7)

Chapter 3 Establishing the Genetic Manipulation Advisory Group

1 Dr Owen was a member of the Employment Medical Advisory Service which had been transferred from the Department of Employment to the HSE.

2 An additional factor would be the eminence and prestige of the scientists on the committee.

3 Some possible confusion is caused by the Working Party's initial adoption of containment categories similar to those in the Godber report, and to their use of related symbols. It would be quite possible that an experimental organism would be placed in a higher category under Williams because of the additional supposed risks involved in the use of genetic manipulation techniques. An experiment involving a non-pathogenic organism (and which was therefore not covered by Godber) would be covered by Williams.

4 All these levels involve physical containment procedures.

5 The arguably novel precautions of 'biological containment' (the creation of disabled experimental organisms unable to survive except under special conditions) found less favour with the Williams Working Party than with Ashby or the Asilomar conference.

6 Owen had been involved in discussions with both CBI and TUC representatives.

7 The TUC and ASTMS had wanted representation on the Williams committee, but this had been rejected by the Secretary of State.

8 An emphasis on the need for continual review of the scientific assessment of risks has been a central part of Brenner's approach, and its appearance here may well owe much to its advocacy.

9 GMAG's considerations of the scientific merit clause are discussed in chapter 4.

10 All individuals (as against representatives of organisations) presenting evidence were scientists. A list is given in Appendix 1 of the Williams report.

11 See chapter 2 for a discussion of the ASTMS input to this set of objectives, many of which have been achieved.

12 This contrasts strongly with the crisis in the first year of GMAG's operation over the problem of the appropriate procedures for commercially sensitive proposals.

13 Lower category experiments could be given the go-ahead by the local safety committee, but the details would still have to be sent to the CAS.

14 The Working Party's earlier reservations on this issue may have been a consequence of wishing to avoid commenting on a sphere they felt properly belonged to the HSE. Questions of 'etiquette' in such

situations loom large in Whitehall, though it must be said that they are also routinely used to defend inaction or indecision.

15 Shirley Williams took a close personal interest in GMAG and was undoubtedly in favour of the inclusion of representatives of the public interest.

16 For recent examples in the context of the sociology of science see, for example, Mulkay (1981) and Mulkay and Gilbert (1982).

17 Given, of course, the prior decision to treat this as a puzzle within the technological paradigm.

18 This was often presented during interviews as a problem of *communication*, with the implication that it was solved as soon as scientists from different areas began talking to each other. It is described here as a problem of *understanding* to emphasise that the processes of (a) recognition of a gap in knowledge and (b) of its repair, were constitutively learning processes which involved, amongst other things, changes in the perception of the problem, and which aspects of it were important.

19 'Scientific' in the actor's sense of the term.

20 In the event, members of GMAG followed the table rather than using it as a flexible resource.

21 Subak-Sharpe (personal communication, 1984) has argued that experience of working with highly dangerous organisms was strongly represented on the Williams Working Party. The relative emphasis on physical containment was thus a consequence of their experience of the necessity for stringent forms of such precautions.

22 The distinction made here is analogous to that of 'arithmomorphic' and 'dialectical' in Georgescu-Roegen (1971) and 'restricted' and 'unrestricted' sciences in Pantin (1968). See also Whitley (1974).

23 It was the coherence that was potentially in danger, not the general assessment of the broad rank order of possible hazards.

24 Another recommendation of the Williams Working Party which appears to owe something to external influence was the reference to regulations under the HASAW Act. Several respondents suggested that the very late inclusion of this item was the result of pressure on the Secretary of State for Education and Science from ASTMS and NALGO.

25 This is not to suggest that all such feedback was necessarily accepted.

26 To be sure, some individual members of the Working Party ceased to have that formal channel of communication and potential influence, but representatives of the various interests continued to have access to the now less visible arena of negotiation.

27 This is a generalisation. For example, as noted earlier, there was a potential divergence between full-time trade union officials and scientist members of ASTMS. Also, although there is less evidence from interviews concerning the later stance of the CBI and the Association of the British Pharmaceutical Industry, it is doubtful that these views had yet been endorsed in any formal sense.

28 DPAG actually operated a temporary system of voluntary control pending the introduction of more formal controls.

29 The ministry responsible for the HSC/HSE is the Department of Employment. The other major possibility that had been canvassed was that the DHSS should be responsible for, or control GMAG, as it already did DPAG. The DHSS had, however, already indicated at the Oxford conference that it did not intend to make a bid for GMAG. It is also possible that there was a minority trade union view favouring the DHSS over the DES, on the grounds that genetic manipulation techniques raised issues that were the proper concern of a department dealing with public health rather than education and science. Some of those interviewed said that the DHSS felt that it had its 'hands full' with the problems of DPAG. One further relevant factor that should be noted here is that ASTMS had developed a rather antagonistic attitude to the DHSS. One person interviewed, well placed to assess, stated that it was the 'stubbornness' of the Secretaries of the ARC and MRC which prevented GMAG from coming under the aegis of the HSE in the early days. The Ministry of Agriculture, Fisheries and Food was also considered, but was never a major contender. A justification of the eventual situation, in terms of DES responsibility for university-based research, was later made by Shirley Williams to a Select Committee of the House of Commons (Select Committee, 1979:153–4 para. 638).

30 In practice the HSC is also responsible for the health and safety of the general public, a point made in its Consultative Document discussed below.

31 The HSC/HSE has a duty to maintain a register of certain possibly hazardous processes.

32 The scientists' response was, in part, organised by the MRC. The definition of genetic manipulation was also criticised, especially by ASTMS, in evidence to the TUC on the HSC Consultative Document and the Williams report in October 1976. In all, close to 100 replies to its Consultative Document were received by the HSC.

33 It has been suggested that the Consultative Document originated within the Inspectorate side of the HSC, who did not have access to appropriate technical knowledge of the area. The latter point was, of course, the burden of argument from a number of scientists. The implication here was that access to such technical knowledge would have avoided the fiasco. While this may well have been true, several points need to be borne in mind. First, it has to be recognised that it was very difficult to produce an adequate definition of genetic manipulation covering all the potentially hazardous new techniques, whilst excluding biological techniques which had been used for many years without substantial concern. Several scientists, and bodies concerned with the area, did in fact try to produce a workable definition, but none of them seem to have been successful.

At one point, the chief scientist to the Central Policy Review Staff of the Cabinet Office (the government 'think-tank') was asked to help the HSE out of its predicament over the definition. Eventually it fell to GMAG itself to produce a definition, a process which took well over a year. Second, although scientists cannot but have been aware of the difficulties of producing a definition, this does not seem to have moderated the criticisms of technical incompetence in any way. Third, the HSE argued that the definition was just as it should be (McDonald, 1976; Lewin, 1976b). It may well have been the intention of the HSE to cover the traditional tools of genetics. Thus Weiss argued at the 1975 Oxford conference concerning the hazards of routine processes (see Weiss, 1975) that there was apprehension that cell fusion techniques might be included in genetic manipulation restrictions, and there were informal pressures to avoid focusing attention on these techniques. Hence questions of competence were hardly the sole consideration in the scientific reaction to the Consultative Document.

34 It appears that the HSE had wanted to settle the question of what kind of experiments GMAG should cover before it began its work. GMAG eventually went ahead on the basis of the formally imprecise definition of the Williams report, and as a consequence had some latitude in deciding what fell within its remit.

35 GMAG's terms of reference and related matters are dealt with in more detail below.

36 This term is used as a convenient shorthand to cover both government departments and other official bodies.

37 Crown property such as hospitals is normally exempt from such regulations, but Ministry of Defence establishments are not. The exemption applies to many Acts of Parliament, not just to the HASAW Act.

38 As noted earlier, the ASTMS policy had included, *inter alia*: adoption of the NIH guidelines, immediate authorisation of some categories of experiment, increased government funding of research in the area, and the establishment of local safety committees.

39 It seems that the dispute about the number of trade union representatives went as far as the Secretary of State. See note 41 below.

40 The AUT was a relative newcomer to the TUC.

41 A messenger from the TUC appeared at the beginning of the meeting to state that, in the opinion of the TUC, the meeting should not proceed, because the 'representatives of the interests of employees' were not present. The chairman relented only to the extent of agreeing that any decisions taken would be discussed at the following meeting. Subsequently the December 1976 meeting became known in the annals of GMAG as the 'Preliminary' meeting, whilst that in January 1977 was referred to as the 'First' meeting. Participants have given several accounts of the reasons for the delay in finalising the employees' representatives: one mentioned TUC

antipathy towards the AUT, but located the delay in the lengthy
bureaucratic consultation procedures of the TUC; another that the
original TUC nominees were not acceptable to the Minister.
According to the *New Scientist* (Lewin, 1976b), the four
nominations from the TUC were not initially accepted by the DES
because of the lack of an AUT member. The AUT pressed its case
on both the TUC and the DES, and also argued that the first planned
meeting of GMAG, in December 1976, should be delayed, but they
were not successful.

42 Further information on this is to be found in Select Committee,
1979: 155 para. 643, and Appendix 15, p. 245.

43 'I would say rightaway that I would certainly not for a moment fail
to consult those whom I knew in the scientific community, quite apart
from the advice given to me by my own Department' (Shirley
Williams, Select Committee, 1979:156 para. 643). One member of
the scientific community well known to the Secretary of State was
Brenner.

44 'I myself believe that a strictly representative committee would be
a very grave mistake and would represent vested interests rather than
representing the best scientific and public knowledge that we have
in a field of this kind' (Shirley Williams, Select Committee,
1979:156 para. 643). It is noticeable that there was only one Fellow
of the Royal Society on GMAG at its inception.

45 See her comments to the House of Commons Select Committee on
Science and Technology. In the context of questions about other
members of GMAG she replied: 'I would regard it as contemptible
if any Minister listened only to a list of names put up by the Civil
Service. I can say at first hand, because I made the appointments
myself . . .' (Select Committee, 1979:155 para. 643). There was a
widespread belief among those interviewed that Shirley Williams
was directly responsible for a number of significant aspects of
GMAG. Indeed, GMAG was often referred to as 'Shirley Williams's
experiment in social democracy in science'. Available evidence
confirms this belief in general, but the inclusion in the Williams
report of a reference to representatives of the public interest was
made whilst Fred Mulley was Secretary of State for Education and
Science, as noted earlier. Shirley Williams took over in September
1976.

46 A counter-argument was articulated in the *New Scientist* editorial
under the title 'Why the haste?' (Anon, 1976b). The Williams
report was criticised for not offering any explanation of why there
was any hurry in allowing experimentation to proceed. It is also
worth noting that early in its life GMAG expressed surprise at the
small number of applications to carry out genetic manipulation
experiments received from scientists. It may be that the immediate
urgency was not as great as was suggested.

47 That is, it was primarily a question of contingent congruence of
interests.

48 It should also be remembered that in any case there was a potential tension within ASTMS between the scientist members' interest in the rapid exploitation of the new techniques, and that of full-time officials who saw the need for more formal systems which clearly established and defined the trade union role. Other objectives, such as representation on local safety committees, were pursued on GMAG itself.

49 The ESF is a body composed of the research councils from sixteen European nations. The proposal had come from its *Ad Hoc* Committee on Recombinant DNA. In September 1976 the Standing Advisory Committee of EMBO (European Molecular Biology Organisation, a scientists' body) had recommended that no attempt should be made to mix aspects of the NIH and Williams systems.

50 One consequence of this was to further entrench the importance of current scientific knowledge in the assessment of possible hazards and appropriate precautions.

51 Dr Haines of the HSE team had assisted the Ashby Working Party at various points.

52 One problem in identifying individuals in this way is a tendency for the analysis to be seen as overly individualistic. The intention here is rather to illustrate the depth of background knowledge about policy making on genetic manipulation that was available to GMAG.

53 Two of the scientists had been members of the Ashby Working Party, and one a member of the Williams Working Party.

54 In the event, the knowledge-base amongst the scientific and medical experts was insufficiently broad. The scientific knowledge of trade union members of GMAG was important in filling the gap. A large number of scientists was also co-opted on to GMAG subcommittees dealing with particular technical issues.

55 The risks were not necessarily thought to be specific to genetic manipulation techniques. For most scientists this was still an open question.

56 This was the prevailing scientific consensus at the time of GMAG's inception.

57 There was no attempt to reduce the level of precautions. As will be seen, the Williams categorisations were quite rigidly enforced.

58 These comments on the centrality of scientific knowledge should not be taken to mean that non-experts were left out in the cold. First, there were a number of important issues which did not require scientific expertise. Second, as familiarity with the Williams system for the categorisation of experiments increased, non-scientist members of GMAG became able in varying part to make their own judgments about the appropriate risk category. That is, to the extent that categorisation became routinised, non-experts could contribute. The importance of the possession of detailed scientific knowledge increased, however, during the consideration of the

implications of the new risk assessment scheme which replaced the Williams system. This is dealt with more fully in chapter 5.

59 Later, the Microbiological Research Establishment was transferred from Ministry of Defence control, and Ellwood could not therefore continue his membership of IPCS. He joined ASTMS, giving them three of the four trade union positions on GMAG.

60 The charge, in fact, follows from the trade unionists' objectives, and from the traditional trade union perception of the need to institutionalise their position and role, rather than from holding separate meetings *per se.*

61 Maddox rejoined *Nature* as Editor in 1979.

Chapter 4 Operating the regulations

1 This was announced by Mr Peter Brooke, Junior Education Minister, in January 1984. GMAG was dissolved and reconstituted as the Advisory Committee on Genetic Manipulation, and as part of the Health and Safety Executive. Some members of GMAG serve on the new committee, but there are no public interest representatives as such.

2 It seems that the DES had specifically wanted this style of chairmanship. Ravetz, a public interest representative on GMAG, stated to the Select Committee that 'GMAG is the least polarised committee on almost all questions that I have ever seen' (Select Committee, 1979:22 para. 71).

3 Of course, one purpose of the Williams Working Party was 'to draft a code of practice and to make recommendations for the establishment of a central advisory service' (Williams Report, 1976:3), and thus to provide policy for the area. However, GMAG's remit was much wider than the purely technical one of the Williams committee, and GMAG was by no means formally constrained to follow the Williams recommendations. For further details see 'The scope of GMAG's operations' below.

4 This is explored in more detail in chapter 6. For the moment it can be noted that the following exchange took place between Arthur Palmer MP, chairman of the House of Commons Select Committee on Science and Technology, and Professor Richmond, a scientist on GMAG:

> *Palmer:* Suppose that we went out to try to find someone representing the public interest and captured that man who is often referred to by high court judges, the man sitting in the Clapham omnibus: he might come and be quite interested in serving, but he would be a bit lost, would he not?
> *Richmond:* He would be totally lost.
> *Palmer:* That is the difficulty, is it not?
> *Richmond:* I think that is the nub of the matter. (Select Committee, 1979:24 paras 76 and 77)

5 Scientists planning experiments were doing so on the basis that they

would be subject to the requirements of the Williams report. They
had been given informal advice on this by the MRC.

6 It has been argued that it was not possible to predict how well the
system would work, and it was necessary to err on the side of caution.

7 How many scientists were waiting to go ahead is a matter of
conjecture. At an early meeting GMAG expressed its collective
surprise at the low number of proposals it had received, and asked
that the press release announcing the establishment of the Group
be circulated to laboratories as a reminder. In research related to
this project which surveyed laboratories which have undertaken
genetic manipulation experimentation, one of the questions
concerned delays caused by the 'Berg moratorium' and subsequent
events. Results indicated that only a few laboratories experienced
any significant delay, but that these were the most important in
the field.

8 There were four levels of containment, I to IV, with IV being the
most stringent.

9 One reason for this was that a licensing system was appropriate only
where hazards were well known and could be assessed in advance,
as was the case with dangerous pathogens. The assessment of the
possible hazards of recombinant DNA techniques, on the other
hand, depended much more on tacit criteria and required a
consideration of the particular details and conditions of the
experiment in question. In practice, repeat experiments, and
experiments which incorporated minor variations, could be
performed without further notification. In more technical terms, the
inserted nucleic acid sequence and the host-vector system had to
be specified in detail on proposals, and could not be changed. In
GMAG's second term, notification procedures for the great majority
of experiments were substantially simplified.

10 Donna Haber (ASTMS) became GMAG's expert on the subject.

11 These problems tended to occur early in the first term of GMAG,
and where a laboratory was first registering its safety committee
with GMAG. As time progressed laboratories entering the area of
genetic manipulation research would presumably already be aware
of GMAG's stance on these issues.

12 For example, whereas the Williams report had allowed that
experiments categorised as I or II by the local safety committee could
be proceeded with at once, GMAG required that scientists await
advice from the Group. This was changed in 1979.

13 In March 1979, Shirley Williams defended the average two- to three-
month delay between GMAG's receipt of an application and the
tendering of its advice as 'by the standards of any committee of this
kind, pretty rapid' (Select Committee, 1979:156 para. 645). A
memorandum submitted to the Select Committee by the Association
of the British Pharmaceutical Industry stated that GMAG had been
'remarkably successful in many respects', and, in particular, that

'GMAG is notably free from bureaucracy and handles proposals with commendable speed' (Select Committee, 1979:197–8).

14 Claims about the relative disadvantage need to be treated with a certain amount of caution, especially in view of the fact that similar complaints were occasionally made by US scientists about their position *vis-à-vis* the UK (and other) guidelines. The two systems were only similar, not identical in terms of the classification of experiments and the levels of precautions. For example, the NIH system, unlike that of GMAG, specified a set of experiments which could not be performed under any circumstances. Again, in 1978, two American researchers from the prestigious Cold Spring Harbor Laboratory travelled to London to perform an experiment under GMAG category III conditions, because no American facility was able to offer the higher containment required by the NIH (Watson and Tooze, 1981:306).

15 As already noted, scientists on GMAG were formally there as experts rather than as members of an interest group.

16 The friction between scientists and GMAG might have been contained, even in the absence of any change in the Group's categorisation system, but for events in the United States, which gave an extra twist to scientists' concerns, and provided an added impetus for GMAG to reconsider its categorisation policy. In the latter part of 1978, the American Recombinant DNA Advisory Committee was in the process of reassessing its categorisation system, and this was widely expected to lead to significantly lower levels of containment.

17 It became difficult to find scientists with a technical knowledge of the area who did not hold consultancies.

18 The majority of the members of the subcommittee were co-opted industrialists. The membership is listed in full in the three GMAG reports.

19 Submission of research proposals was, of course, voluntary pending regulations under the HASAW Act. In addition, GMAG's pronouncements only had the status of advice, even though this might be considered to be definitive in the eventuality of litigation arising from any actual harm. This was never tested in the courts. Thus it would have been possible for industry to, for example, notify the HSE of their intention to perform unspecified experiments and state that they would not accept any information being passed on to GMAG.

20 In this early suggestion it would be up to the proposing scientist to decide whether the proposal should be dealt with by this special procedure.

21 This seems a rather formal requirement designed to emphasise the gravity of concern. However rigid the legal requirement, it would probably have been extremely difficult to enforce in practice.

22 It does not seem to have been made clear whether this threat would apply to all of those formally representing the interests of

management, to all GMAG members with industrial links, or just to those who were fully employed by industrial concerns.

23 One difficulty was that whatever the current position, British patent law was being redrafted as part of a move to bring it into line with proposed pan-European standards.

24 Like all of GMAG's subcommittees, meetings were open to all full members of the Group. This one was apparently well attended by a wide range of the Group's membership.

25 At one point Wolstenholme met representatives of the ABPI for an exchange of views.

26 The analogy found little favour with the majority of GMAG members. One point of difference was that the other bodies mentioned did not consider detailed technical information of the type used by GMAG.

27 This is the manner in which the trade union suggestion was portrayed during interviews. The industrialists felt that they had already changed their position to a significant degree, and in order to meet trade union objections.

28 The draft was apparently prepared by the chairman in consultation with a scientist, an industrialist and a trade unionist.

29 Under this proposal the trade unionists would not even know if a confidential submission had been considered, and thus would not be able to consult with their members in the relevant laboratory.

30 This might have risked the withdrawal or resignation of the trade unionists. A modification to the mini-GMAG idea, which would have allowed any member of the Group to see, but not to take part in discussions of, any confidential proposal, though only at the express agreement of the proposer, was also rejected by the trade unionists.

31 Under this scheme, only those with potentially conflicting interests or connections would be prevented from examining proposals, whereas the mini-GMAG option excluded a number of people without commercial links.

32 Apparently no proposals were considered before February 1978.

33 It seems that the chairman, the secretary, one scientist, one industrialist and one trade unionist worked together to create this draft.

34 Trade union or employee members of the local genetic manipulation safety committee would, of course, be covered by the firm's own confidentiality arrangements.

35 One of the trade unionists, Owen, wanted to retain his freedom to discuss matters with people outside GMAG, and therefore did not sign. One other GMAG member, a scientist, felt unable to sign the confidentiality agreement for personal reasons. Both these individuals had, of course, to withdraw when confidential proposals were discussed.

36 A Select Committee of the House of Commons made a series of criticisms about the special confidentiality scheme. In addition to matters of general principle, they were concerned that the

withdrawal of GMAG members with technical knowledge could mean that very few, or none, of those remaining had any direct experience of the techniques (Select Committee, 1979, and Anon, 1979b and c).

37 This is not to suggest that the public interest representatives, for example, did not take part in the debate, but rather to emphasise that the trade unions made the running, and had the necessary 'clout'.

38 It was emphasised in interviews that personal integrity was not in doubt; the problem was perceived as one of inadvertent passing on of information.

39 The rather clubby nature of GMAG and the associated stress on individual honour and trust seem to have aided this.

40 One example here would be the development of a bacterium which, in symbiosis with cereal crops, would fix nitrogen from the air thus reducing the need for expensive nitrogenous fertilisers.

41 This was in large part due to the experimental difficulties encountered in attempting to introduce foreign DNA into plant tissues.

42 'Plant pests' include, of course, members of the animal as well as the plant kingdom.

43 Any experiments north of the border would be dealt with by the Department of Agriculture and Fisheries for Scotland.

44 Nor did the HASAW Act cover the use of the products of genetic manipulation.

45 The environmental implications of the eventual release of genetically engineered organisms were sidestepped. GMAG had discussions with MAFF and with the Natural Environment Research Council on the matter, but the issues turned out to be highly complex. The expertise required to make even the roughest of assessments about the ecological effects was simply not available. It was not until the publication of GMAG's third report that the environmental implications were commented on, and then it was only to state that the Group should be informed of experimental work leading to release at the earliest stage, and that it would deal with such proposals on a case-by-case basis (GMAG Third Report, 1982:10). To summarise, in the absence of the availability of detailed technical knowledge which could be brought to bear on the matter, GMAG would necessarily have to rely on the tacit knowledge of scientists, as in the earlier stages of the genetic manipulation debate.

46 Advice notes 1 and 6; these are reprinted in GMAG's First Report (1978:42 and 53).

47 Contra-indications included: the use of immunosuppressive drugs, taking antibiotics also used in the experiment, defective barriers to infection such as disorders of the skin, lungs or alimentary canal, and, more generally, any physical or psychological unsuitability. Women who were, or who expected to become, pregnant were

advised to consider whether or not they should become involved with certain classes of experimental work.

48 The conditions of 'good microbiological practice' are given in the GMAG Third Report (1982:123). These guidelines were produced by the Joint Co-ordinating Committee for the Implementation of Safe Practices in Microbiology.

49 Other subcommittees dealt with the risks of scale-up, the requirements for high containment facilities, the validation of safe vectors, and the criteria for new guidelines.

50 A related point was made by a member of the Select Committee who challenged a senior civil servant to agree that GMAG's relations were 'a cat's-cradle, really, of Ministerial responsibility . . . and that, on the face of it, there must be a case in the near future for rationalising these lines of responsibility . . .' (Select Committee, 1979:114 para. 440).

51 The smooth passage of these regulations through the parliamentary process was threatened when a group of Labour MPs lodged an objection on the grounds that they failed 'to give the necessary surveillance to safeguard the population from the accidental creation of organisms against which the human body would have no defence . . .' (Morris, 1978). In effect, the MPs were about four years behind the current scientific consensus. The regulations were also criticised by the MPs for failing to control the products of genetic manipulation experiments after they had left the laboratory.

52 There were also problems with the definition of genetic manipulation which had been formulated for the HSE by GMAG. These are dealt with in the second part of this chapter.

53 The trade unions, in particular, doubted the effectiveness of the Act in the light of such phrases. See the comments by Jenkins and Sherman, 1977:117.

54 Experimental work was restricted to the somewhat arbitrary limit of ten litres of culture. A subcommittee was in the process of dealing with this barrier in the context of work involving the scale-up towards production.

55 Following negotiations within the TUC there was a further, unsuccessful, attempt to gain a fifth trade union seat on GMAG. The official version of the negotiations is that when the TUC came to consider renominations for the second term it recognised the justification of the AUT's case for a place. The AUT had a larger number of affected members than other affiliated unions. The TUC took the view that not to renominate would imply criticism of, or lack of confidence in, the present incumbents. As what looks rather like a placatory gesture to the AUT, it was decided to renominate the four incumbents, but to ask for the fifth seat. The issue does not seem to have been pressed too hard.

56 A similar view is expressed by McClelland (1978). However, the criteria on which decisions about the continued membership of public interest representatives were taken remains obscure.

57 Similar sentiments can be found in Henderson's statements to the Select Committee, 1979:35 *et seq*. Henderson was, from the beginning of 1979, deputy chairman of the Working Party on Biotechnology, which was jointly set up by the Royal Society, the Advisory Board for the Research Councils, and the Advisory Council for Applied Research and Development. The Report of the Working Party contained criticisms of GMAG's regulations in terms of their having inhibited the scientific and industrial exploitation of genetic manipulation (Spinks, 1980).

58 It must be assumed that the future needs of the Group, and the particular views of Henderson, were taken into account by those deciding on the new chairman.

59 The members of the Technical Panel were chosen on the basis of their technical expertise in genetic manipulation. They were Ellwood, one of the trade unionists, and three of the scientific/ medical expert members of the Group: Langley, Walker and Wildy. Additional members could be co-opted as necessary. Walker, the chairman of the Technical Panel, had joined the Group at the beginning of 1979. He had served on the Williams Working Party, and was one of the scientists responsible for the Williams system of categorisation. Later, the Technical Panel was replaced by the Technical Subcommittee which had a different membership.

60 As noted in the Introduction, a considerable body of work in the sociology of science has shown that the radical separation of the scientific and the social is not epistemologically tenable, and that there are social processes involved in even the so-called pure sciences.

61 The HSE was by now engaged in its own extensive consultation procedures on the new risk assessment scheme. This involved consulting some sixty bodies.

62 The self-cloning issue had been a bone of contention at the December 1978 meeting of GMAG and scientists, at which the new risk assessment scheme had been introduced.

63 The publication of the note was an urgent matter for GMAG in view of the impending international conference at Wye College, Kent, organised by COGENE, the Committee on Genetic Experimentation of the International Council of Scientific Unions, a scientific rather than a trade union body. The significance of the conference for GMAG's actions over the note on self-cloning was that it did not want to be the butt of criticism on this point; the note was published just a few days before the conference.

64 It is estimated that two-thirds of all proposed experiments fell into category I or GMP.

65 As an index, the Group met seven times in 1980 and in 1981 only five times, whereas before it had met monthly.

66 A year earlier the Technical Panel had been asked to examine the NIH regulations to see whether there was anything worth adopting.

67 There were now only three public interest representatives. Maddox

had resigned in the early part of 1980 on taking up the editorship of *Nature*, arguably the world's leading scientific journal. Although there were later changes to the membership of GMAG, the public interest representation remained at three.

68 GMAG's third report shows only one subcommittee, the Technical Subcommittee. Its membership, which would have varied according to the number of co-opted members, is given, for February 1982, as twelve. The chairman was Walker; there were three trade union members, five co-opted scientists (i.e. not GMAG members), and one each of the following: public interest representative, scientific member of GMAG, assessor (HSE).

69 By asking for annual returns, GMAG could use its remit to monitor genetic manipulation work. In effect, GMAG had introduced a licensing system, an idea initially floated by the public interest representative, Maddox, in 1978, and rejected by the Group.

70 The process of postponing difficult decisions for future consideration was, of course, a recurrent feature of GMAG's deliberations. In this case, the delay may well have been to allow the trade unionists to consult their constituency. Another comment relevant at this point is that by now the trade unionists' practice of holding their own discussions before GMAG meetings was perceived by some to be of benefit to the work of the Group. It was suggested in interviews that this served the purpose of enabling the scientist trade unionists to persuade their non-scientist colleagues of the need for various changes. This observation seems to have some validity.

71 The other experiments included category III and IV experiments, any which involved the production of more than ten litres of culture, and any case where the local safety committee was unable to reach a decision.

72 The HSE had some reservations about the details of what was essentially a code of practice for GMP. Since the publication of a guidance note (note 14) was urgent, it went ahead anyway and consideration of the HSE's points put off for later consideration. In fact, the HSE had been producing a draft code of practice to cover such work, and this had now been superseded by the issue of note 14. It was estimated by Henderson that in the future only about 2 per cent of planned experiments would have to await GMAG's advice (see Anon, 1980a). Later, an *ad hoc* committee under Walker met to produce a further version, which was published as note 15 in January 1980. Almost immediately, this had to be revised after sharp protests from scientists, who found that the details of the code of practice represented a significant raising of precautions. This seems to have been a mistake on GMAG's part; it had no intention of tightening the regulations.

73 See McKie (1979) for an account of an interview with Carlisle, then Opposition spokesman on Education and Science.

74 Other items on the agenda included the abandoning of formal approval procedures and the circulation of proposals.

75 Subak-Sharpe had been the chairman of the subcommittee for the validation of safe vectors.

76 This perspective was expressed in Anon (1981), an article obviously based on discussions with a participant.

77 In probably the most comprehensive experiment, over a period of twenty-six months in Bristol, and twelve months in London and in Seattle, there was no evidence among sixty-four laboratory workers nor their families that they acquired strains of *E. coli* even when conjugative, readily transmissible plasmids were used and work was carried out with no special precautions but using normal microbiological techniques (Petrocheilou and Richmond, 1977). Walgate (1979) cited Richmond, a scientist on GMAG, to the effect that he had carried out some risk assessment experiments 'to show willing', and that if GMAG wanted such research done 'it may have to force them [scientists] to do it'. Another researcher is quoted as saying that 'We have run out of good ideas for risk assessment experiments'.

78 The figure is quoted in an editorial in *Nature* (Anon, 1978).

79 There were four meetings in 1981, three in 1982, none in 1983, and the final meeting in January 1984.

80 It was accepted that this was an important issue, but it was felt that it was difficult to formulate advice for 'conjectural' hazards, to plants for example, at that stage. The Natural Environment Research Council was studying the possible impact of genetically altered organisms in forests. It was agreed to keep the topic of environmental implications under review.

81 There was continuing interest, within the Council of Europe, in the topic of human rights in relation to possible therapeutic applications of genetic manipulation. Gibson, the Secretary of GMAG, attended several meetings including an *ad hoc* committee which, in 1983, planned a series of discussions on moral, ethical and legal aspects.

82 This view was also put to the Select Committee of the House of Commons on Education, Science and the Arts during their enquiry into biotechnology, in May 1982, where it was argued, as part of the ASTMS submission, that GMAG should have its remit broadened so that it could consider ethical and safety aspects of biotechnology.

83 During 1982 the Department of Industry conducted a study of the impact of health and safety legislation and the regulatory environment on biotechnology. The exercise may have been prompted by the Spinks report (1980) on biotechnology which called on GMAG to do nothing which would hinder Britain's competitive position. (The Group had, throughout, denied causing any significant delays to industry, and members in interviews have cited the case of a company choosing to carry out research in this country because of the stable regulatory environment.)

84 There had been a feeling within the MRC that the officials of some departments were procrastinating. At one point it was suggested

that Brenner should talk to the departmental representatives on GMAG so that they could relay the need for action.

85 In practice this would probably have meant the trade unions making it a major issue, and one which would potentially threaten their long-term goal of control of the area by the HSE.

86 At this point there were three public interest representatives.

87 There was a fairly common perception that there was little of a specific nature that could be identified as their unique contribution *qua* public interest representatives, except the valuable symbolic function they had performed, especially in the early days.

88 The Report listed four options which, in summary, were:

(a) to continue the current arrangements,
(b) to have two separate advisory committees, one reporting to Ministers, the other to the Health and Safety Commission (HSC),
(c) to reconstitute GMAG along the lines of the Advisory Committee on Dangerous Pathogens, with a joint DES/HSC secretariat,
(d) to reconstitute GMAG as an advisory committee to the HSC.

89 An application was already under consideration in the United States.

90 The Department of Health and Social Security, in 1985, has the Working Party on Recombinant DNA of the Standing Advisory Committee studying these issues.

91 An informal meeting of government officials, and including some members of GMAG, had already been held.

92 See, for example, Bartels (1983) and Bartels *et al.* (1983).

Chapter 5 Reconceptualising hazards and risks

1 Kuhn (1970), and recent work in the sociology of science, provide a series of examples of scientists using what is now known (or believed) to be true to reinterpret earlier events. See Weiner (1979) for an historian's corrective to this form of rational reconstruction.

2 One fear voiced in the early days was that the important scientific advances to be made might attract 'cowboys' who would have little regard for sensible precautions.

3 See, for example, the comments in paragraphs 1.5 and 5.4 of the Williams report. Arguably, the report might have given greater prominence to the need for flexibility. On the other hand, GMAG's terms of reference stated that it should 'undertake a continuing assessment of risks and precautions' (GMAG First Report, 1978:1).

4 Brenner was, however, unsuccessful in gaining acceptance for his view that the detailed guidelines were too inflexible.

5 The 'proposition' was submitted before any clear GMAG approach had been developed.

6 Brenner had hoped that the Safe Vectors Subcommittee would agree to this first part of the paper being sent to GMAG as a 'sampler', and that a specialist committee would be set up to develop the ideas

further. In the event, the more lengthy process of consideration detailed below was adopted.

7 A later, combined, version of the three parts is reprinted in GMAG's Second Report (1979:77–90).

8 Scientists now find it difficult to reconstruct the rationale for this approach. It may well be that the belief 'the closer the phylogenetic relationship, the greater the likelihood of hazard' derived from the inherent nature of biological knowledge, based, as it is, on evolutionary theory. Thus, during the early development of the Williams scheme and the discussions leading to the Asilomar conference and NIH guidelines, and without experimental information and fundamental consideration, the degree of evolutionary relatedness was assumed to be of great importance.

9 A point essentially identical to this had been contained in the 'proposition' from Brenner's laboratory to GMAG in early 1977.

10 A similar wording is to be found in the revised version of Brenner's paper (GMAG Second Report, 1979:77–90).

11 A similar mode of analysis, based on a consideration of the sequence of events necessary to produce a hazardous outcome, was outlined in the Report of the First Meeting of the Standing Advisory Committee on Recombinant DNA of EMBO, held in February 1976 (reproduced in Watson and Tooze, 1981:213) and, apparently in a simpler form, in evidence to a public hearing of an advisory committee to the Director of the NIH, also in February 1976 (Krimsky, 1982:173–4). A subsequent and more detailed analysis in the same mode, including the specification of approximate probability values for various events, and concluding that the hazards were vanishingly small, was published in the *New Scientist* (Holliday, 1977; reproduced in Watson and Tooze, 1981:215–17).

12 A similar analysis of the rhetorical devices used in discussions of radiation hazards can be developed. See, for example, Mazur, 1973, who looked at both the radiation and the fluoridation controversies.

13 Ravetz had raised the question of attempting some form of risk assessment at a GMAG meeting in the summer of 1977, but had met with no support (see chapter 4).

14 Several of those interviewed argued that Rigby's contribution was crucial to the success of the scheme.

15 Apparently Brenner helped to create some momentum by contacting Wolstenholme, the chairman of GMAG.

16 Maddox had hoped that the new risk assessment scheme would lead to a greater effective relaxation of categorisation than the version which was finally accepted.

17 Quotations have been taken from the published version, which does not differ greatly from the later unpublished versions.

18 It may be stretching the Kuhnian model too much to suggest that, on a small scale, the change in the basis of categorisation resembled a mini-scientific revolution, with Brenner's paper providing the necessary exemplar for a 'gestalt switch' to take place. The

Williams approach had generated enough anomalies to create a
potential crisis of confidence, but it is also clear that extra-scientific
pressures (for example, the possible NIH guideline changes, a 'brain
drain', etc.) played a key role in changing the basis for
categorisation.

19 In reply to those who argued that the whole business of
categorisation was 'unscientific', it could be argued that this ignored
the 'gut feeling' of scientists that some experiments were more
hazardous than others.

20 This was not uniform because some canonical/paradigm cases were
known and others were not. Also increasing knowledge led to slightly
enhanced ability to provide figures.

Chapter 6 Science and public participation

1 Krimsky (1982:14) also links this with social unrest stemming from
the rise of black civil rights movements in the USA, which together
formed the 'twin impacts of the abrupt tearing of the American
social fabric'. Also see, *inter alia*, Turney (1977) and Dale (1981).

2 For early, but cogent, critiques of Bell's post-industrial society thesis,
see Kumar (1976 and 1978). There is some doubt about what
exactly constitutes a professional or technical employee, since, for
example in Chicago, a garbage collector is labelled a 'sanitary
engineer'. Kumar suggests that relabelling may involve a
'sociological sleight of hand' which significantly undermines the
official figures. A more recent reassessment of the whole thesis can
be found in Gershuny and Miles (1983).

3 It is instructive to note the wording of the title of Davis's (1977)
article in the *American Scientist*: 'The recombinant DNA scenarios:
Andromeda strain, Chimera and Golem.'

4 Indeed, many scientists were both surprised and dismayed at the
strength of the public reaction.

5 An important example is the report of the Ashby Working Party
(Ashby, 1975).

6 Wright (1978), Yoxen (1979a) and Krimsky (1982) have analysed
the reasons why the moratorium was taken seriously and the
regulations, by and large, adhered to by the scientists.

7 Mazur (1973) has also considered disputes between experts and
lawyers.

8 Krimsky notes that 'Public participation in the rDNA policy process
had its debut, at least in a symbolic sense, at the meeting held on
9–10 February 1976 of the NIH Director's Advisory Committee'
(1982:169). This public advisory committee, quite separate from
the RAC, has broad membership and advises the NIH on the wider
aspects of biomedical science policy. At this point the RAC
contained two non-scientists.

The American debate was characterised by a much greater
degree of openness than in Britain. In addition to the various local

initiatives such as the Cambridge 'citizens' court' described below, there were a series of Congressional committees and hearings at which scientists and their arguments were subjected to public scrutiny. Finally, local Institutional Biosafety Committees, whose remit covered genetic manipulation, contained 20 per cent of members drawn from the community at large.

9 Although the incident did not involve the use of genetic manipulation techniques it raised general fears about the safety of many areas of possibly hazardous biological research.

10 To be fair, there was little or no relevant experience which could be drawn upon in considering the mechanics of public interest representation on GMAG.

11 The extent to which information was made available to lay members of the RAC is not known.

12 This became increasingly difficult in the later phases of GMAG, when there was less controversy and public concern over the possibility of hazard.

13 There can be no doubt that the scientists tried to be helpful. The first chairman of GMAG, Wolstenholme, has written that:

> No praise could be too high for the manner in which our scientists made the principles of their esoteric techniques painstakingly comprehensible, even to those with absolutely no education in science; equally, the lay members patiently strove to understand the points that mattered in regard to risk, safety and health. (Wolstenholme, 1984:13)

However, the comments made in interviews indicate a somewhat less rosy picture.

14 The scheme also had early backing from another public interest representative, Ravetz, who had some experience of risk analysis techniques.

15 In practice, the possibilities immanent in the American system seem to have been partly offset by the 'encyclopaedic' nature of the NIH guidelines, and its consequent emphasis on detailed knowledge of a large range of experimental parameters.

16 Having a constituency, and an effective means of canvassing opinions about, and support for, policies was important on GMAG. Indeed, the Genetic Manipulation Working Group of the Confederation of British Industry was created at the instigation of one of the representatives of the interests of employers to fulfil just these needs.

17 This stands in sharp contrast to the American experience with the Cambridge Experimentation Review Board, where the committee was made up of members with equal, low scientific status and expertise, but with the real power to discontinue work should this be deemed necessary. However, as Krimsky (1978 and 1982) has pointed out, the American system also has its faults, not the least of which is the rather simplistic division of the dispute into two opposing camps. Another is the very lengthy procedure required to arrive at a

decision with, in this particular instance, over 100 hours of testimony and deliberation. Krimsky (1982:310) concluded that a typical result of local responses to the DNA controversy was 'cosmetic' branches on the prevailing rules.

18 Effective performance of the role of juror relies on the presence of adversary debate to open up the issues on which judgment may be made. This was facilitated by the American system. On GMAG, because of the closed nature of the debate, the juror also had to act as counsel for both defence and prosecution to properly examine the evidence.

Chapter 7 Conclusion

1 Discussion of individual proposals, necessary for the establishment of case law, would in any case had to have been held in confidence.
2 The exceptions were its paper on the new risk assessment scheme, and evidence given to the Select Committee.
3 Maddox, a public interest representative, did at one point attempt to initiate public discussions on the possibility of GMAG adopting a system of licensing laboratories to conduct genetic manipulation research for a fixed period rather than dealing with individual proposals.
4 A series of training courses for biological safety officers, held at the Microbiological Research Establishment, Porton Down, and required by GMAG, were also influential in ensuring good safety standards.
5 The direct cost of GMAG to the government was very low. A number of scientists did however spend a good deal of time devising and operating the regulations, and others time complying with them.

Bibliography

Anon (1975), 'Forever amber on manipulating DNA molecules?', *Nature 256*:155.

Anon (1976a), 'Why are we waiting?', *Nature 262*:244.

Anon (1976b), 'Why the haste?', *New Scientist* 2 September 1976:474.

Anon (1978), 'Now reason can prevail', *Nature 276*:103.

Anon (1979a), 'The voice of the people – which one?', *Nature 278*:675.

Anon (1979b), 'GMAG secrecy worries MPs', *New Scientist* 25 January 1979:236.

Anon (1979c), 'Select Committee questions GMAG's membership', *New Scientist* 15 March 1979:845.

Anon (1980a), 'Still looser UK guidelines', *Nature 287*:265–6.

Anon (1980b), 'Will the DNA guidelines wither away?', *Nature 287*: 473–4.

Anon (1981), 'Genetic engineering: watchdog lives to bite again', *New Scientist* 1 January 1981:7.

Ashburner, M. (1976), 'An open letter to the Health and Safety Executive', *Nature 264*:2–3.

Ashby (1975), *Report of the Working Party on the Experimental Manipulation of the Genetic Composition of Micro-organisms*, London: HMSO (Cmnd 5880).

Barnes, B. (1974), *Scientific Knowledge and Sociological Theory*, London: Routledge & Kegan Paul.

Barnes, B. (1977), *Interests and the Growth of Knowledge*, London: Routledge & Kegan Paul.

Barnes, B. and Edge, D. O. (1982), *Science in Context*, Milton Keynes: Open University Press.

Bartels, D. (1983), 'Oncogenes: implications for the safety of recombinant DNA work', *Search 14(3–4)*:88–92.

Bartels, D., Naora, H. and Sibatani, A. (1983), 'Oncogenes, processed genes and safety of genetic manipulation', *Trends in Biochemical Sciences 8(3):*78–80.

Bell, D. (1973), *The Coming of Post-Industrial Society*, New York and London: Heinemann.

Bennett, D. J., Glasner, P. E. and Travis, G. D. L. (1984), *Report on the Development and Operation of Genetic Manipulation Regulations in Britain*, London: Polytechnic of North London.

Berg, P., Baltimore, D., Boyer, H. W., Cohen, S. N., Davis, R. W., Hogness, D. S., Nathans, D., Roblin, R., Watson, J. D., Weissman, S. and Zinder, N. D. (1974), 'Potential biohazards of recombinant DNA molecules', *Science 185:*303; *Proc.Nat.Acad.Sci.US 71:*2593–4.

Berg, P., Baltimore, D., Brenner, S., Roblin III, R. O. and Singer, M. F. (1975), 'Summary statement of the Asilomar conference on recombinant DNA molecules', *Science 188:*991–4; *Nature 255:*442–4; *Proc.Nat.Acad.Sci.US 72:*1981–4.

Bishop, J. O., Murray, K., Murray, N. E., Walker, P. M. B. and Williamson, R. (1974), 'Statement to the Ashby Working Party', unpublished document.

Brenner, S. (1974), 'Evidence for the Ashby Working Party', unpublished document, Laboratory of Molecular Biology, Cambridge, UK.

Brenner, S. (1975), Transcript of interview between S. Brenner and C. Weiner, 21 May 1975. Recombinant DNA Oral History Collection (MC 100), Institute Archives and Special Collections, MIT Libraries, Cambridge, Mass.

Brenner, S. (1978), 'Six months in category four', *Nature 276:*2–4.

Brenner, S. (1981), Interview transcript.

British Society for Social Responsibility in Science (1978), 'The politics of genetic engineering', in *Science for People*, 39:1–4.

British Society for Social Responsibility in Science (1979), 'Memorandum submitted by the Genetic Engineering Group of the British Society for Social Responsibility in Science', in *Select Committee* (1979); 206–16.

Chargaff, E. (1977), quoted in Grobstein, C., 'The recombinant-DNA debate', in *Recombinant DNA, Scientific American*, San Francisco: W. H. Freeman.

Cockburn, C. (1981), 'Bug business is big business: cloning for capitalism', *Science for People 48:*1–3.

Cohen, S. (1973), *Folk Devils and Moral Panics*, London: Paladin.

Collins, C. H., Hartley, E. G. and Pilsworth, R. (1974), *The Prevention of Laboratory Acquired Infection*, Monograph no. 6, London: Public Health Laboratory Service.

Collins, H. M. (1975), 'The seven sexes: a study in the sociology of a phenomenon, or the replication of experiments in physics', *Sociology 9:*205–24.

Collins, H. M. (1981), 'Stages in the empirical programme of relativism', *Social Studies of Science 11:*3–10.

Collins, H. M. and Pinch, T. J. (1982), *Frames of Meaning: The Social Construction of Extraordinary Science*, London: Routledge & Kegan Paul.

Cox (1974), *Report of the Committee of Inquiry into the Smallpox Outbreak in London in March and April 1973*, London: HMSO (Cmnd 5626).

Dale, A. M. (1981), *The Recombinant DNA Debate: A Documentary Analysis*, unpublished MSc thesis, University of Surrey.

Davis, B. D. (1970), 'Prospects for genetic intervention in man', *Science* 170:1279–83.

Davis, B. D. (1977), 'The recombinant DNA scenarios: Andromeda strain, Chimera and Golem', *American Scientist LXV*, September/October:547–55.

DHEW (NIH) (1976), *Recombinant DNA Research. Vol. I. Documents relating to 'NIH Guidelines for research involving recombinant DNA molecules'*, February 1975–June 1976, DHEW publication no. (NIH) 76-1138, Bethesda, Maryland: NIH.

Douglas, M. (1975), *Implicit Meanings*, London: Routledge & Kegan Paul.

Edge, D. O. and Mulkay, M. J. (1976), *Astronomy Transformed*, New York: Wiley Interscience.

Federal Register (1976), 'Recombinant DNA research guidelines', *Federal Register 41*:27907–43.

Federal Register (1977), 'Recombinant DNA research – proposed revised guidelines', *Federal Register 42*:49596–609.

Fischhoff, B., Slovic, P. and Lichtenstein, S. (1981), 'The public vs. "The experts": perceived vs. actual disagreements about risks', in Kunreuther, H. (ed.), *Risk: A Seminar Series*, Laxenburg, Austria: International Institute for Applied Systems Analysis.

Ford, B. T. (1974), 'Call for biohazard legislation', *Nature 250*:364–5.

Galbraith, J. K. (1969), *The New Industrial State*, Harmondsworth: Penguin.

Georgescu-Roegen, N. (1971), *The Entropy Law and the Economic Process*, Cambridge, Mass.: Harvard University Press.

Gershuny, J. I. and Miles, I. D. (1983), *The New Service Economy: The Transformation of Employment in Industrial Societies*, London: Frances Pinter.

Gilbert, G. N. and Mulkay, M. J. (1984), *Opening Pandora's Box: A Sociological Analysis of Scientists' Discourse*, Cambridge: Cambridge University Press.

GMAG (1978), 'Genetic manipulation: new guidelines for UK', *Nature* 276:104–8.

GMAG First Report (1978), *First Report of the Genetic Manipulation Advisory Group*, London: HMSO (Cmnd 7215).

GMAG Second Report (1979), *Second Report of the Genetic Manipulation Advisory Group*, London: HMSO (Cmnd 7785).

GMAG Third Report (1982), *Third Report of the Genetic Manipulation Advisory Group*, London: HMSO (Cmnd 8665).

Bibliography

Godber (1975), *Report of the Working Party on the Laboratory Use of Dangerous Pathogens*, London: HMSO (Cmnd 6054).

Grimwade, S. (1977), 'Recombinant DNA', *Nature 270*:291.

Habermas, J. (1968), *Knowledge and Human Interests* (trans. Shapiro, J. J. (1971)), London: Heinemann.

Habermas, J. (1976), *Legitimation Crisis* (trans McCarthy, T.), London: Heinemann.

Hanson, E. D. (ed.) (1983), *Recombinant DNA Research and the Human Prospect*, Washington, DC: American Chemical Society.

Hesse, M. (1980), *Revolutions and Reconstructions in the Philosophy of Science*, Brighton, Sussex: Harvester Press.

Hirsch, W. (1962), 'The image of the scientist in science fiction: a content analysis', in Barber, B. and Hirsch, W. (eds), *The Sociology of Science*, New York: Free Press.

Holliday, R. (1977), 'Should genetic engineers be contained?', *New Scientist* 17 February 1977:339–401.

Hollis, M. and Lukes, S. (1982), *Rationality and Relativism*, Oxford: Basil Blackwell.

HSC (1976), *Compulsory Notification of Proposed Experiments in the Genetic Manipulation of Micro-organisms*, London: HSC.

HSC (1978), *Health and Safety (Genetic Manipulation) Regulations*, London: Statutory Instrument 1978 no. 752.

Jenkins, C. and Sherman, B. (1977), *Collective Bargaining*, London: Routledge & Kegan Paul.

Jenkins, C. and Sherman, B. (1979), *White Collar Unionism: The Rebellious Salariat*, London: Routledge & Kegan Paul.

Johnston, R. (1976), 'Contextual knowledge: a model for the overthrow of the internal/external dichotomy in science', *Australian and New Zealand Journal of Sociology* 12:193–203.

Johnston, R. (1980), 'The characteristics of risk assessment research', in Conrad, J. (ed.), *Society, Technology and Risk Assessment*, New York: Academic Press: 105–22.

Krimsky, S. (1978), 'A citizen court in the recombinant DNA debate', *Bulletin of the Atomic Scientists* 34(8):37–43.

Krimsky, S. (1979), 'Regulating recombinant DNA research', in Nelkin, D. (ed.), *Controversy: Politics of Technical Decisions*, Beverly Hills: Sage.

Krimsky, S. (1982), *Genetic Alchemy: The Social History of the Recombinant DNA Controversy*, Cambridge, Mass: MIT Press.

Kuhn, T. S. (1970), *The Structure of Scientific Revolutions*, Chicago: University of Chicago Press.

Kumar, K. (1976), 'Industrialism and Post-Industrialism: Reflections on a putative transition', *Sociological Review*, new series 24(3):439–78.

Kumar, K. (1978), *Prophecy and Progress*, Harmondsworth: Penguin.

Lakoff, S. A. (1977), 'Scientists, technologists and political power', in Spiegel-Rösing, I. and Price, D. J. de S. (eds), *Science, Technology and Society*, London: Sage.

Latour, B. and Woolgar, S. (1979), *Laboratory Life: The Social Construction of Scientific Facts*, Beverly Hills: Sage.

Lewin, R. (1976a), 'Genetic engineering and the law', *New Scientist* 28 October 1976:220–1.

Lewin, R. (1976b), 'No workers' representatives yet for GMAG', *New Scientist* 16 December 1976:635.

Lewin, R. (1977), 'GMAG falls foul of privacy constraints', *New Scientist* 15 December 1977:683.

Locke, J. (1976), 'An open reply from the Director of the Executive', *Nature 264*:3.

Maddox, J. (1972), *The Doomsday Syndrome*, London: Macmillan.

Majone, G. (1979), 'Process and outcome in regulatory decision-making', *American Behavioural Scientist 22*(5):561–83.

Mazur, A. (1973), 'Disputes between experts', *Minerva XI*(2):243–62.

McClelland, A. J. (1978), 'Blind man's bluff at GMAG', *Nature 276*:657–8.

McDonald, J. A. W. (1976), 'Genetic engineering', *New Scientist* 11 November 1976:353.

McKie, R. (1979), 'Britain's Shadow Science Minister believes in experts', *Nature 278*:387.

Medawar, P. B. (1963), 'Is the scientific paper a fraud?', *The Listener* 12 September 1963:377–8.

Medawar, P. B. (1977), 'The DNA Scare: Fear and DNA', *New York Review of Books*, 27 October 1977.

Merton, R. K. (1942), 'Science and technology in a democratic order', *Journal of Legal and Political Science 1*:115–26.

Merton, R. K. and Barber, E. (1963), 'Sociological ambivalence', in Tiryakian, E. A. (ed.), *Sociological Theory, Values and Sociocultural Change: Essays in Honour of Pitrim A. Sorokin*, 91–120. New York: Free Press.

Mitroff, I. I. (1974), 'Norms and counter-norms in a select group of the Apollo moon scientists: a case study of the ambivalence of scientists', *American Sociological Review 39*:579–95.

Morgan, J. and Whelan, W. J. (eds) (1979), *Recombinant DNA and Genetic Experimentation*, London: Pergamon Press.

Morris, M. (1978), 'Genetic engineering code "not tight enough"', *Guardian* 28 June 1978.

MRC (1975), Press release, Department of Education and Science: London.

Mulkay, M. J. (1979), *Science and the Sociology of Knowledge*, London: Allen & Unwin.

Mulkay, M. J. (1980), 'Sociology of science in the West', in Mulkay, M. J. and Milic, V., 'The sociology of science in East and West', *Current Sociology 28*(3):1–184.

Mulkay, M. J. (1981), 'Action and belief or scientific discourse?', *Philosophy of the Social Sciences 11*:163–71.

Mulkay, M. J. and Gilbert, G. N. (1982), 'What is the ultimate question? Some remarks in defence of the analysis of scientific discourse', *Social Studies of Science 12*:309–19.

Nelkin, D. (1977a), *Technological Decisions and Democracy: European Experiments in Public Participation*, Beverly Hills: Sage.

Nelkin, D. (1977b), 'Technology and public policy', in Spiegel-Rösing, I. and Price, D. J. de S. (eds), *Science, Technology and Society*, London: Sage.

Nelkin, D. (1978), 'Threats and promises: negotiating the control of research', *Daedalus 107*(2):191–209.

Nelkin, D. (1979), 'Science, technology and political conflict: analysing the issues', in Nelkin, D. (ed.), *Controversy: Politics of Technical Decisions*, Beverly Hills: Sage.

NIH (1977), *Environmental Impact Statement on NIH Guidelines for Research Involving Recombinant DNA Molecules, Parts 1 and 2, DHEW Publication (NIH) 1489 and 1490*, Bethesda, Maryland: NIH.

Pantin, C. F. A. (1968), *The Relations between the Sciences*, Cambridge: Cambridge University Press.

Petrocheilou, V. and Richmond, M. H. (1977), 'Absence of plasmid or *Escherichia coli* K–12 infection among laboratory personnel engaged in r-plasmid research', *Gene 2*:323–7.

Piel, G. (1979), 'Scientific research: determining the limits', in Wulff, K. M. (ed.), *Regulation of Scientific Enquiry: Societal Concerns with Research*, AAAS Selected Symposium 37, Boulder, Col.: Westview Press.

Polanyi, M. (1958), *Personal Knowledge: Towards a Post Critical Philosophy*, London: Harper & Row.

Price, D. J. de S. (1963), *Little Science, Big Science*, New York: Columbia University Press.

Ravetz, J. R. (1973), *Scientific Knowledge and its Social Problems*, Harmondsworth: Penguin.

Ravetz, J. R. (1979), 'DNA-research as "high intensity" science', *Trends in Biochemical Sciences 4*:N97–N98.

Rogers, M. (1975), 'The Pandora's Box congress', *Rolling Stone* 19 June 1975:37–40, 42, 74, 77–8, 82.

Rogers, M. (1977), *Biohazard*, New York: Knopf.

Rose, H. and Rose, S. (1969), *Science and Society*, Harmondsworth: Allen Lane.

Rose, H. and Rose, S. (1976), 'The radicalisation of science', in Rose, H. and Rose, S. (eds), *The Radicalisation of Science*, London: Macmillan.

Royal Society (1975), 'Comments by the Royal Society on the Report of the Working Party on the Laboratory Use of Dangerous Pathogens (Godber report, 1975)', Press release 14 July 1975, London: Royal Society.

Royal Society (1979), 'Flaws in GMAG's guidelines', *Nature 277*:509–10.

Sandbach, F. (1978), 'The rise and fall of the Limits to Growth debate', *Social Studies of Science 8*:495–520.